CATALYST MANUFACTURE

Second Edition

CHEMICAL INDUSTRIES

A Series of Reference Books and Textbooks

Consulting Editor

HEINZ HEINEMANN
Berkeley, California

ADDITIONAL VOLUMES IN PREPARATION

CATALYST MANUFACTURE

Second Edition

Alvin B. Stiles

University of Delaware
Newark, Delaware

Theodore A. Koch

E. I. du Pont de Nemours & Company
Wilmington, Delaware

University of Delaware
Newark, Delaware

CRC Press
Taylor & Francis Group
Boca Raton London New York

CRC Press is an imprint of the
Taylor & Francis Group, an **informa** business

CRC Press
Taylor & Francis Group
6000 Broken Sound Parkway NW, Suite 300
Boca Raton, FL 33487-2742

First issued in paperback 2019

© 1995 by Taylor & Francis Group, LLC
CRC Press is an imprint of Taylor & Francis Group, an Informa business

No claim to original U.S. Government works

ISBN-13: 978-0-8247-9430-9 (hbk)
ISBN-13: 978-0-367-40178-8 (pbk)

**Visit the Taylor & Francis Web site at
http://www.taylorandfrancis.com**

**and the CRC Press Web site at
http://www.crcpress.com**

Preface

As I look back at the first edition, which was published a decade ago, I find that much of catalysis has remained the same, particularly with respect to secrecy. However, with respect to the catalysts, their operation and composition, there have in some cases been a great number of changes – in other cases the old standbys are still the old standbys. I do not belong to the old school that says, "Let well enough alone," but I believe that even in the cases in which there have not been a great number of changes and insignificant improvement, there still is room for research. I urge anybody with such an inclination to do research despite the fact that he or she may encounter from some quarters the statement, "This has been the standard for 50 (or however many) years, and the likelihood of making an improvement is very remote." I thoroughly disagree, both for personal reasons and because I know of cases in which such a statement has proved to be misleading and unwisely discouraging.

This second edition specifically addresses the changes that have been made in zeolites, which are high-temperature oxidation catalysts for total oxidation, particularly in conjunction with gas-fired turbines. The text gives much consideration to environmental catalysts and the problems related to the environment. There is some consideration of a unique CO oxidation catalyst about which there has been substantial publicity, but which has seen little adaptation to the commercial scale. We all recognize methane coupling as a subject that has been intensely considered for the synthesis of acetylene, ethylene, and aliphatic homologs of ethylene. In those cases in which catalyst manufacturing was described in detail in the first edition, both as to procedure and to equipment used, an effort has been made to bring the discussion up to date.

As one considers the environmental catalysts, one must also consider the shifting thoughts as to what constitutes the environment. "Environment" is much more broadly interpreted by government and other agencies than is the current practice. For example, it may include living conditions in the home, dietary choices, smoking and drinking habits, insect life, pesticides, and combinations of things such as the foregoing and clothing. These, for the most part, do not have anything to do with the environment we consider in speaking of catalysis. It is readily evident that this broader area of "environment" includes items that cannot be altered by simple catalytic means. Since we are consciously excluding enzymes, some of the foregoing environmental items drop out. However, one of the more important areas of environmental catalysis is the abatement of automotive fumes, and this subject is brought up to date and extended into the future by facts and, where indicated, appropriate speculation.

Areas of very great effort in the zeolite area concern the so-called phosphates and activated carbons derived from chlorinated polymers. Although they are termed zeolites, these are actually microporous solids that have zeolitic characteristics in that they selectively absorb molecules of a certain size and configuration. The phosphates, I believe, are being used in the industry at only one site, and one of the drawbacks of the phosphates is that they are reportedly lower in stability than the silica aluminas.

Another very important subject considered in this edition is the utilization, recovery, and regeneration of used catalysts. This subject is of special importance to those selling catalysts or initiating a new catalytic process. Many of the components of catalysts are either water-soluble or weather to a soluble form, such as when chromites oxidize to chromates. Disposal and recovery methods vary widely, and it is our intention to describe the many alternatives.

Still another major change in the past ten years has to do with the safety requirements for various pieces of equipment. In the earlier edition many pieces of equipment were described; the manufacturers have provided us with information about alterations they have made in the equipment to comply with OSHA and this has been added appropriately in this edition.

Catalyst manufacture is one of the most interesting areas in which one can engage. It is characterized by having a great number of surprises and unique conditions that can be discovered only by much trial and error and, unfortunately, many false starts. I have now spent nearly sixty years in this discipline, and I have enjoyed every minute of it; I have always felt that many of the discoveries were derived not from intelligence alone but from intelligence guided by good luck and intuition.

I would like to extend special thanks to Theodore A. Koch, whose hard work and diligence made this second edition possible.

Alvin B. Stiles

Introduction

The second edition of *Catalyst Manufacture* has two primary purposes. The first is to bring up to date the current manufacturing procedure for those catalysts that have enjoyed significant beneficial changes in composition or preparation procedures. The second is to incorporate into the discussion of each catalyst manufacturing, use, discharge, and disposal procedure instructions or recommendations to make these procedures legal in light of EPA and OSHA regulations.

The EPA regulations concern the dust, aqueous discharges, organic solvents, and other contaminants that would be adsorbed by the catalyst during use or formed during oxidation or partial oxidations when exposed to air. The used catalyst itself, whether contaminated or not, also presents a challenge for disposal, regeneration, or recovery of metals values. In this book, each of the foregoing is considered in the light of each member of each family of catalysts. Furthermore, a listing of companies specializing in recovery of metals values is given in the Appendix. Regeneration and reuse schemes are provided where appropriate; however, there is frequently a limit to the number of times a catalyst can be reused, and there is often the need to screen to remove fines that must be reprocessed by one or more of the procedures described.

Numerous regulations introduced by OSHA relate to the equipment and procedures used for manufacturing catalysts as well as discharging, storing, regenerating, and reusing or recovering the used catalyst. Inasmuch as the used catalyst is nonuniform from time to time, it must, as a general principle, be assumed to be coated with toxic materials.

In the equipment chapters, special attention is given to the need for special clothing, ventilation, and equipment guards. Much of this was not required when the first edition was published (but was very likely practiced due to corporate regulations). Governmental regulations and requirements are included herein to the extent that the application is known or can be visualized by the authors. Hazards resulting from slippery floors, tripping obstructions, or hot surfaces are considered here only in a general way but are, of course, noted by government inspectors.

One realm of regulation that was previously not of critical concern is the noise level of the operation. Examples of equipment used in catalyst manufacture that must be observed for this problem include grinders, crushers, ball mills, granulation mixers, tableting machines, and coaters.

The regeneration, reoxidation, or oxidation of catalysts for discharge may produce carbon monoxide, aldehydes, or, at times, unidentified toxic discharges. Catalytic combustion can usually be used to eliminate these by-products, but not if they include catalyst poisons such as halides, ash, sulfur, and solids that deposit on and blind the catalyst. These conditions may entail retaining a professional disposal company; such companies usually advertise in scientific or commercial journals and may even be listed in the Yellow Pages of the telephone directory.

Infractions of government regulations may entail substantial penalties of several types. As a consequence it behooves the business management to be familiar with their content and purposes. We hope that the foregoing and this book will help in avoiding costly problems.

Contents

Part I
Equipment

1
Scope and Goals

The goal of this book is to present a lucid and accurate picture of catalyst preparation on a laboratory, pilot plant, and commercial scale. A further purpose is to present the pertinent available information, whether it is published, in the open literature, or in patents, without infringing upon any proprietary information. If the information is in unexpired patents, this is mentioned in the text. If the information is in expired patents that are in the public domain, it is not specifically identified as such. It is possible that certain features of the descriptions given herein are patented and have patent matter relating to them of which we are unaware. Therefore it is wise to make a patent search before engaging in commercial practice of any process. In many cases, catalyst manufacture is not patented because the owners prefer to keep the process secret and operate as a secret shop rather than disclosing the information in a patent that will eventually become public

domain or, even more likely, will make the information fair
game for those who are skilled in avoiding the art as taught.

It is a further goal of this work to give information that
will make it possible for someone skilled in catalytic science
to prepare a given catalyst on either a laboratory, semiworks,
or commercial scale. It is, of course, impossible to give information for all types of catalysts, but it is our intent to give the
procedures for all those that are comparatively well known to
the practicing catalytic scientist.

In the preparation of many catalysts, there are few, if any,
problems in increasing the quantity prepared from that typical
of the laboratory scale, that is, 25–50-gram quantities, to semiworks scale (usually from a few liters to a 100 or so liters per
day) or to the scale of a commercial plant that would be producing ton quantities per day. However, by contrast, it is also
occasionally true that the increase in size of production is extremely difficult to accomplish. This may be the case when
there are problems relating to

1. The temperature of precipitation
2. The rate of precipitation
3. Agitation type and degree
4. Aging effects of the precipitate
5. The effect of washing and soluble salt removal, occasionally
   influenced by the particle size of the precipitate, which in
   turn is affected by the aforementioned factors

and finally,

6. Drying rate and uniformity of calcining conditions

An additional factor is that on a small scale it is typically
impossible or very difficult to prepare the catalyst in pelleted
or extruded form, which parallels the conditions employed in
a commercial operation. Pilling, pelleting, granulation, and
extrusion all have a significant effect on catalyst activity. It
is recognized, of course, that the catalyst must be put into a
physical form that will permit its use in gas, liquid, or slurry

services, and in order to accomplish this it is necessary that the typically finely divided powders be converted to some granular, spherical, or cylindrical form.

This operation must be accomplished with the minimum adverse effect on the catalyst. A point that must be made at this time is that when a catalyst is prepared and used on a small scale and is not put through one of the densification and particle-forming operations, that catalyst may perform entirely differently than if it were subjected to the particle-forming operation. It is even misleading to pellet a catalyst as one would for plant operation and then crush and screen it and use it in the granular form for a small-scale test, because some of the adverse effects of the densification or particle-forming operation are obviated by the crushing and screening. Consequently, in such a case it should always be noted in test data that the catalyst that has been used has been pelleted but has been further modified by the crushing and screening operation.

A further consideration of the effect discussed in the preceding paragraph can be roughly quantified by recognizing that a catalyst that is used in powder form, such as nickel on kieselguhr, and is not pelleted is, of course, not subject to the performance deterioration that accompanies the particle-forming procedures. If, however, a catalyst is prepared as for acrylonitrile synthesis, that is, by spray-drying the ultimately prepared mixture, the adverse effect in this case is relatively minor but still notable.

A particulate type with a slightly more severe adverse effect is encountered when the catalyst is converted to a "mud-like cake" by compression or kneading and is then dried and converted into a ceramic-like "brick," which is then crushed and screened to form granules. (This procedure is roughly what is followed for iron molybdate type methanol oxidation catalust, activated aluminas, activated carbon, and some dehydration catalysts.) An operation that has a yet greater adverse effect is for the catalyst in a paste form to be extruded through a barrel-type extruder. This requires no pilling lubricant and

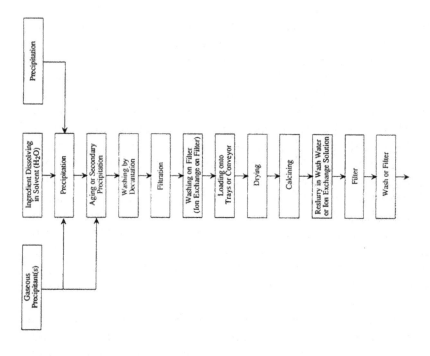

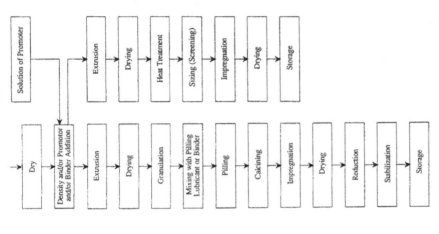

Figure 1 Block flow diagram of all-inclusive procedure for manufacture of precipitated catalysts.

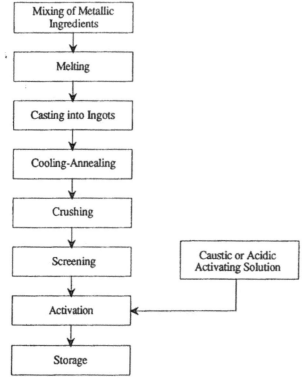

Figure 2 Block flow diagram of basic procedure for preparation of fused metallic ingredients. Example: nickel aluminum alloy.

results in a relatively hard extrudate that on drying becomes sufficiently hard in some cases for commercial operation. This catalyst is only slightly more adversely affected than one subjected to the previously described granulation technique.

The final operation to be considered in this scale of "adverse effect" is the pilling operation. This is usually accomplished by placing a powder into a tabletting machine fitted with dies and punches so that the powder is compressed into cylindrical forms. Other tablet forms, of course, are possible and will be described more fully later, but the typical form is

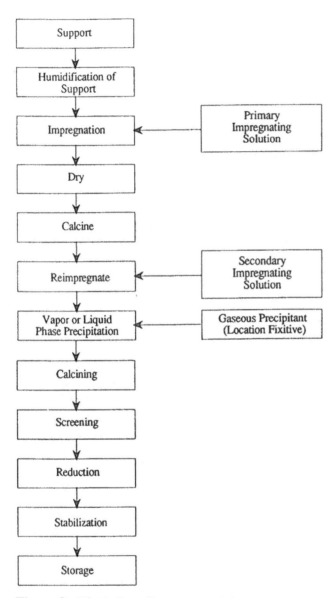

Figure 3 Block flow diagram of all inclusive procedure for the preparation of impregnated catalysts.

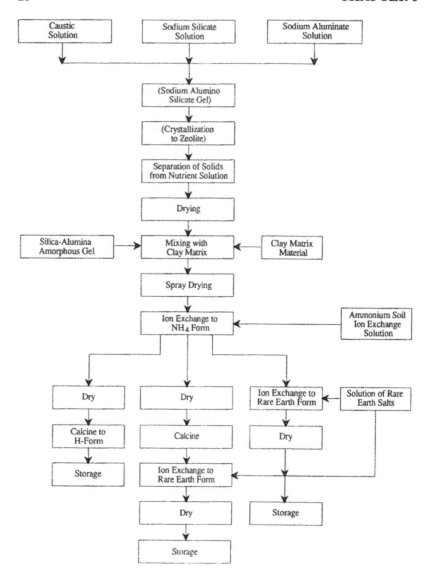

Figure 4   Block flow diagram of basic procedure for the preparation of zeolites in acid and rare earth form.

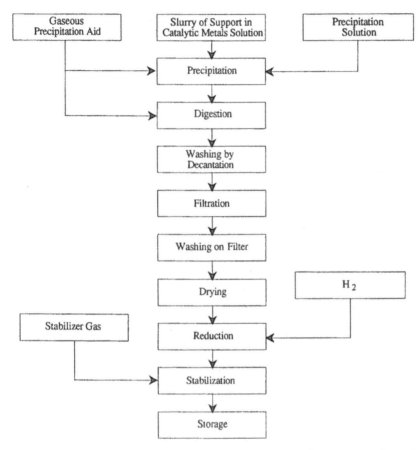

Figure 5 Block flow diagram of procedure for the preparation of reduced and stabilized metal on powdered support. Example: Nickel or copper on kieselguhr.

that of a right cylinder with length and diameter essentially equal. One of the major problems in the pelleting operation arises from the need to add a lubricant such as stearyl alcohol, polyvinyl alcohol, polyvinyl acetate, graphite, talc, or grease. These problems will also be described more fully later. A secondary factor is that in many cases, in order to redevelop the

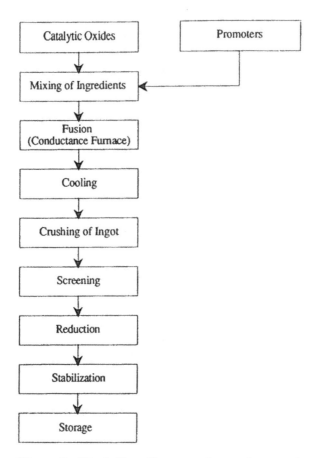

Figure 6   Block flow diagram of procedure applicable to the preparation of fused oxide catalysts. Example: Fused promoted iron oxide for ammonia and Fischer-Tropsch synthesis.

closed pores, the catalyst is exposed to an oxidation operation to remove the combustible lubricants. In this operation, the oxidant is intended to be atmospheric oxygen, but the oxygen may be derived by reducing one or more of the catalyst ingredients, thereby producing a deactivating exotherm or compositional alteration. The conversion of the catalyst to particulate structures is a very sensitive operation and will be described more fully subsequently.

Many catalysts are used in the form of impregnated particles. This introduces many additional factors that must be considered — for example, the strength of the support, its reactivity with the catalyst ingredients, its pore size and pore distribution, blockage of the pores of the support with catalytic material, epitaxial effect of the support in altering the crystallinity or atomic or crystallite spacing of the catalytic ingredients added thereon, and finally locating the catalytic materials on the support material in such a position as to make them available to the reactants of the process (see Fig. 7). In further consideration of this latter factor is the fact that the catalytic materials are usually purposely applied and positioned exclusively to the peripheral surface. This means that abrasion of that surface due to any type of movement in a catalyst bed may abrade away the catalytic materials, leaving only a bare uncoated smaller support. It should be added also that in many cases in which a reactor is described as "fixed bed," experience will show that there actually is substantial motion

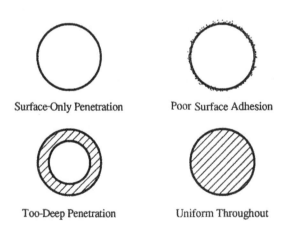

Surface-Only Penetration   Poor Surface Adhesion

Too-Deep Penetration   Uniform Throughout

Figure 7   Diagrammatic representation of four examples of the results of typical impregnation conditions.

and shifting of the catalyst bed. This can be detected in many ways such as by placing the catalyst in layers and then observing the alteration of those layers after the catalyst is removed from the converter. It frequently is quite startling to observe the degree to which mixing has occurred; a concomitant event to this movement is the gravitating of fines to a location where they cause pluggage.

Alloy catalysts, such as nickel aluminum alloy, will also be discussed, and their use in both powdered and granular form will be described. Activation of the alloy will be described for both types of forms, and the intricacies of the activation and problems that are apt to be encountered will be mentioned, with advice on how to avoid these problems.

Precious metals catalysts will be discussed, both those in massive, that is, screen form, and those supported on carbon and other types of supports. Some of the advantages and disadvantages of carbon and other supports will be mentioned, and some schemes will be discussed whereby the catalytic materials can be located at previously selected positions, either on the surface or throughout the entire support, according to the preference of the catalytic scientist or the plant operator.

Fused oxide catalysts will also be covered as they are prepared for ammonia synthesis and indirectly for Fischer-Tropsch synthesis.

The preparative procedures are given in block outline form in Figures 1-7, and these figures should be referred to throughout the reading of this book.

2
Catalysts Prepared by Precipitation

The plan of approach for this book is to, first of all, describe each of the preparative requirements for catalysts in general and for preparation on the laboratory, pilot, and commercial scale. Equipment will be considered as to function (i.e., precipitation) on each scale individually and sequentially as to procedure. The composition, fabrication, and activation of various "families" of catalysts (e.g., petroleum processing catalysts) are considered in separate chapters. The modifications that must be introduced into the preparation of individual catalysts within each family to achieve unique properties are presented within the chapter dealing with that family.

Parenthetic to this discussion of precipitation procedures, in many applications, rough screening of candidate catalysts can be made effectively and relatively easily (thus avoiding the more time-consuming precipitation) by choosing an "inert" porous support such as alpha-alumina, mullite, or sillimanite and then supporting the catalytic materials such as metals, oxides, or mixed oxides by placing soluble salts on them and calcining the salts. This, of course, is usually not a preferred way to make the catalyst, but it is quite satisfactory for a rough screening of materials on a go/no-go basis. A more detailed method of preparation of these rather crude supported candidates will subsequently be given under the topic of supported catalysts preparation.

In the preparation of laboratory-scale lots of catalysts, the most frequent method of preparation is via precipitation. A schematic diagram of laboratory facilities is shown in Fig. 8. It can be seen that a simple glass beaker is quite satisfactory

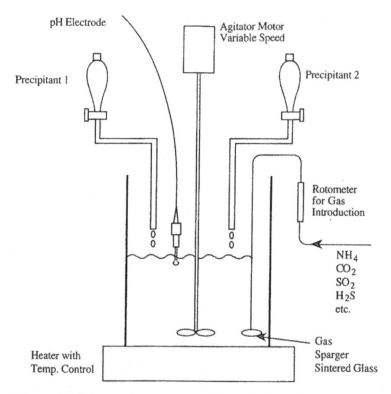

Figure 8 Schematic representation of principles and equipment used in the preparation of precipitated catalysts.

for the precipitation vessel. Dropping funnels or burets are also quite satisfactory for the addition of the precipitant. Temperature is indicated by either a thermocouple or a thermometer. The difficulty with a mercury thermometer is that if for whatever reason the mercury bulb is broken, the mercury contamination may destroy the catalyst. An immersion electrode for pH indication is also a part of the setup.

It should be observed that on occasion some precipitation may occur on the pH electrode, with the result that faulty readings will be obtained depending upon the type of catalyst being precipitated. The corrective measure for dissolving this precipitate should be selected after consideration of the catalyst components. The agitator is ordinarily a glass paddle that is rotated by a variable-speed, relatively low rpm drive. It is best to have the entire setup in a well-ventilated hood to remove ammonia, halides, or other offensive odors. Heating of the beaker usually is accomplished by having the beaker sitting on a electrically heated hot plate with variable wattage input for the temperature control; an electrically heated jacket could also be used. The rate of agitation should be variable, because some catalysts (for example, carbonates) are susceptible to deterioration or decomposition if agitated too vigorously. Other catalysts are susceptible to air introduced by excessively rapid agitation, whereas others produce a dense precipitate that settles out when agitation is insufficient. As a rule of thumb, approximately 200–300 rpm is adequate. The glass agitator is ordinarily set up so that it covers approximately 50–70% of the diameter of the beaker. The agitator also is situated in such a way that it essentially touches the bottom of the beaker, thus avoiding precipitate buildup on the bottom.

In certain conditions it may be necessary to simultaneously add two precipitants or a precipitant and a slurry of the support material. This is accomplished by having a second buret or separatory funnel. It also may be desirable to add a gas simultaneously with the precipitation or even as a precipitant itself. If, for example, anhydrous ammonia is added to effect precipitation or for the purpose of redissolving the precipitate as a metallic ammine, it can be added through a sintered glass sparger located very close to the agitator paddle in order to achieve the greatest amount of agitation and rapid dispersion of the ammonia.

Carbon dioxide may also be added simultaneously with the ammonia for in situ formation of an ammonium carbonate or for other reasons such as to increase the carbonate content of a basic carbonate precipitate derived from sodium bicarbonate. Another feature that is sometimes desirable is to have an interlock between the temperature inside the precipitation vessel and the heating vessel in order to control the temperature more accurately within the precipitation vessel. Such a mechanism is highly desirable during the precipitation stage, when the vessel's contents may be rapidly cooled by a cool liquid precipitant with the result that a large fraction of the precipitation occurs at a temperature other than the optimum.

Although the casualty rate for glass equipment is much greater than that for stainless steel, it is preferred for most catalysts to use glass to avoid contamination of the catalyst with metals dissolved from the stainless steel container. This is not to say that stainless steel is not satisfactory in many operations, but it is always a question as to what the corrosion characteristics are for a new catalyst composition. It is noteworthy that chlorides, acidified tungstates, molybdates, and many of the heteropolyacids are extremely corrosive even to stainless steel.

There are occasions when the precipitation should preferably be conducted under superatmospheric temperature and pressure. This could involve the use of glass-lined pressure equipment (Fig. 9) such as that built by Pfaudler Company. Pfaudler offers equipment of various sizes and materials of construction with various draw-off and addition lines, with jacket and internal temperature control, and various agitator types, preferably generating a gentle movement of the internal liquid or slurry. These elevated temperature and pressure vessels are usually employed in those cases in which the catalyst

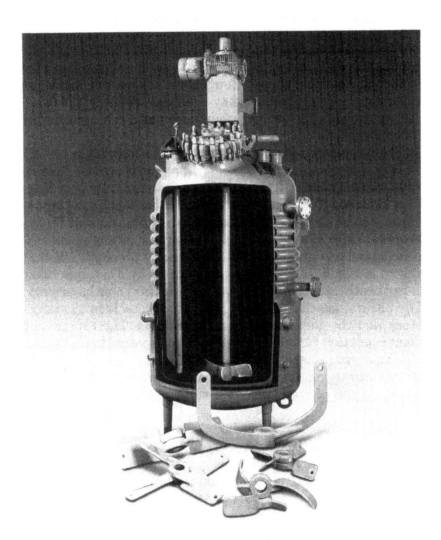

Figure 9 Exterior view of Pfaudler glass-lined tank and examples of the various impellers available for agitation of the contents of the tank. (Courtesy of Pfaudler-US, Inc., 1000 West Avenue, Rochester, NY 14692.)

is desirably modified by a hydrothermal treatment following the precipitation. Superatmospheric conditions may also be applicable when gaseous precipitants such as ammonia and/ or carbon dioxide are used. Supplementary carbon dioxide at superatmospheric pressure may be employed when it is desirable to increase the proportion of carbonate in a basic carbonate precipitate.

In commercial precipitation there is a very wide variety of suitable precipitation vessels. Some of the most reliable types of tanks are still the wooden variety built by New England Tank and Tower. There are many other fabricators whose names can be obtained from the *Chemical Engineering Equipment Guide* or similar sources. The general arrangement of a commercial precipitation system is, except for scale, like that shown in Fig. 8. It can be quickly noted why a catalyst plant is frequently referred to as an overgrown inorganic laboratory. The basic similarity of the precipitation vessels in the laboratory and the commercial plant is evident. The commercial plant's solution tanks for the precipitants very closely resemble the buret or separatory funnel of the laboratory. Its pH, temperature, and flow controls, however, are usually automatic instead of the typically manual controls of the laboratory scale.

Although we have described the wood precipitation equipment and some precipitant storage tanks, it is obvious that in many cases stainless steel is implied for plant use. When stainless steel is used, it is strongly recommended that very careful initial observations and study of corrosion be made. This is good business for two reasons. First, of course, is the fact that if corrosion is severe, the equipment simply will not last. But the most important feature for the catalytic scientist or the operator of the catalytic plant is that the catalyst usually becomes contaminated with any corrosion products that form. If these corrosion products are tolerable, which may be the case, then the problem is not a severe one. On the other hand, in many cases these impurities have a drastically adverse effect on the catalyst. To convert stainless steel to an acceptable material of construction, exposed surfaces can be lined with Teflon or sprayed with polyethylene or polypropylene, or the vessel can be lined with rubber or, in some applications, even plated with a nonreactive metal. All of these options are called upon under certain circumstances, so no fixed rule is advisable at this time.

A frequent problem in plant operations in which the same equipment is used successively for different kinds of catalysts is the cleanup of the equipment between types of catalysts. Usually the tank must be suitable not only for all of the catalytic materials that are made in it but also for certain acid treatments that may be necessary to dissolve catalysts caked on the walls or in interconnecting lines. If heating and cooling are accomplished by a steam or water jacket, the interior surface is comparatively free from obstruction for cleaning by maintenance people entering the tank. If, on the other hand, internal cooling and heating coils are used, it is almost impossible to manually scrub the tank, and cleaning must be done by using acid or another solvent solution. Internal coils have the advantage of providing better and more rapid heat exchange, but cleaning is difficult, and this fact may be the deciding factor in avoiding them.

In many cases when catalysts are being prepared, such as in the preparation of molecular sieves in which crystallization occurs and crystals may grow on the interior walls of the vessel as well as on all other parts of the equipment, it is advisable, insofar as possible, to schedule the equipment so it is used exclusively for a given type of operation. When processing any catalyst that has a strong tendency for cake buildup on the walls or other parts of the vessel, it is inadvisable to attempt to use the precipitation facilities for multicatalyst preparation.

3

Solution and Slurry Transfer

It is often necessary to transfer a solution from one vessel to another by means of a pump (gravity flow should be provided where possible), or to pump slurry from one vessel to another or from the precipitation vessel to filtration facilities. When transferring a solution, there is no problem relating to the shearing of the precipitate, and as a consequence, a high-speed centrifugal pump is quite satisfactory. By contrast, when a slurry of a precipitate or a suspension of a molecular sieve is being transferred or withdrawn from a tank, care must be taken to avoid the structural alteration that can be caused by shearing or abrasion. As a consequence, in pumping slurries, it is necessary to consider other types of pumps such as a reciprocating, diaphragm, or lobe pump or a screw-type pump with rubber housing such as the Moyno pump. The diaphragm and Moyno pumps are pictured in Figs. 10 and 11, respectively. There are many manufacturers of each type of pump, and these are tabulated in the *Chemical Engineering Equipment Guide*.

(A)

(B)

Figure 10 (A) Exterior view of diaphragm pump. (B) Diagrammatic representation of piping scheme for installation of diaphragm pump.

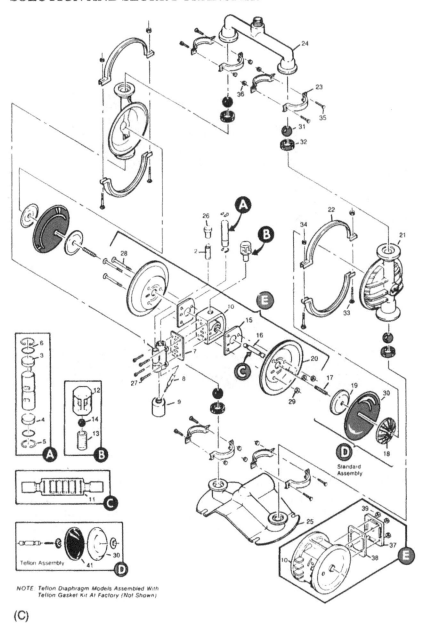

(C)

Figure 10 (C) Exploded view and enlargement of arrangement of components of double-acting diaphragm pump. (A, courtesy of Wilden Pump and Engineering Company, 22069 Van Buren Street, Colton, CA 92324.)

(A)

(B)

Figure 11 (A) Moyno pump, exterior view. (B) Cutaway view of
Moyno pump showing parts and principle of operation. (A, courtesy
of Robbins and Myers, Inc., 1895 W. Jefferson St., Springfield, OH
45501 and Geiger Pump and Equipment Co., 600 Jewell St., Delmar,
DE 19940.)

Materials of construction for pumps can be almost limitless. Each type of material has its advantages and disadvantages. Duriron, for example, has many advantages from the standpoint of resistance to abrasion and corrosion, but it is brittle and therefore susceptible to fracture by heavy-handed maintenance or operating people. Teflon is used, as are polypropylene and polyethylene, for both semiworks and plant-scale equipment. Stainless steel of various types and Hastelloy can also be cast or fabricated into suitable pump housings and impellers.

For a good all-purpose transfer and relatively high pressure and volume discharge pump system, the diaphragm pump is highly satisfactory for use in catalyst manufacture.

4
Filtration

Slurries of catalytic materials are frequently difficult to filter because they are made in such a way as to achieve maximum surface area for the resultant catalyst. This often means, but not always, that the catalyst has gone through a gel stage during precipitation, and, as is well known, gel-like precipitates are frequently very difficult and slow to filter. On the laboratory scale, it is not difficult to compensate for this slow filtration simply by allowing sufficient time for the filtration and for washing on the filter, but in industrial and commercial practice this becomes an expensive factor and occasionally an intensely difficult problem to resolve. A corollary to slow filtration is that the gel-like structure of the precipitate may occlude foreign ions, which must be removed from the catalyst if it is to function at its maximum efficiency.

Problems in filtration sometimes can be avoided or minimized by alterations in the preparation procedure. Occasionally, increasing or lowering the temperature of precipitation will improve filtration; in other cases, increasing or decreasing the rate of precipitation effects an improvement; in still other cases, concentrating the solutions may be beneficial. As a last resort, recycling powdered finished catalyst as a filter aid may be effective in improving filtration. Filtration is frequently a problem also with the finished catalyst when it is used, for instance, in a slurry operation and the finely divided catalyst must be removed from the product. When a problem is encountered at this stage, it is not uncommon that it can be remedied or at least somewhat alleviated by precipitating the catalyst on a support, which may also improve the dispersion of the catalyst. The support may be effective in improving the filtration characteristics of the catalyst both initially and when it is used in its final form in the catalytic operation.

Frequently used support materials are alumina hydrate, kieselguhr (diatomaceous earth), powdered silica, and calcium carbonate as well as others too numerous to mention. Powdered activated carbon is frequently employed because of its valuable and unique properties of acid and base resistance, high surface area, making for excellent catalyst dispersion, and low density, making for easy suspension in a liquid medium.

Laboratory-scale filtration is usually performed using a Buchner funnel with Whatman type filter papers. There are many types of Whatman filter papers, and these, of course, can be selected for satisfactory performance depending upon the filtration characteristics of the precipitate to be processed. It is frequently desirable to use two filter papers to mimimize the chance of the slurry bleeding through the filter media. One of the annoying characteristics of filter paper is its tendency to adhere to the filter cake when the cake is discharged from the Buchner funnel. The difficulty of peeling the paper away from the filter cake can be avoided by using a filter cloth cut to the circle size required for the Buchner funnel being used. Cloth filter is not a complete answer because it sometimes is difficult to seal the edges, and under such circumstances it may be desirable to employ a combination, using paper below the filter cloth next to the Buchner face. In this case the filter cloth acts as a separatory medium between the filter cake and the filter paper. As is the case with filter papers, there are many types of filter cloths, including duck and twill as well as others that are satisfactory and much thinner. The weave can be made quite thin, and the cloth can be woven from many types of fibers such as nylon, Teflon, polyester, glass, or even metallic wires.

The pH and temperature of the slurry are critical in the selection of the filter media. It could be mentioned additionally that insofar as possible it is advisable to have the slurry at a pH slightly below neutral (acidic) because then the filtration and washing of the filter cake are frequently more readily accomplished than in a high pH or basic environment. The higher surface tension of a basic slurry tends to make the filtration slow and washing more difficult.

In extreme cases when only a small sample of catalyst is required, separation of the liquid from the solid can be accomplished by centrifugation. This is rarely practiced in the preparation of commercial quantities. Centrifugation of large batch sizes on a commercial basis is not impossible, but it is ordinarily practiced on crystalline and easily filtered materials (see Fig. 16).

FILTRATION—SEMIWORKS SCALE

As one increases the size of the catalyst preparation from laboratory to semiworks scale, the Buchner type filter is also applicable at the larger scale. Although the general design is much like the Buchner, it is now referred to as a Neutsche filter (Fig. 12). The principle is exactly the same, however; that is, a vacuum chamber beneath the filter medium is used to provide a pressure differential between the slurry and the filtrate, which forces the liquid through the filter medium and effects the retention of the filtered cake on the filter medium. Filter cloths are almost universally used on Neutsche filters, which can be

(A)

Figure 12   (A) Exterior view of Knight-Ware ceramic vacuum filter.

obtained in sizes ranging from a few gallons capacity to as much as several hundred gallons. The larger sizes could more correctly be described as commercial scale.

The next most frequently used filter method is the plate-and-frame system (Fig. 13). These filters can be constructed from wood, hard rubber, polypropylene, Teflon, or a metal such as stainless steel. The frame can be as small as 8 in. or, on special order, even smaller; semiworks sizes may be up to 18 in. The number of plates and frames can be varied to accommodate larger batches, and the plates and frames can be so designed

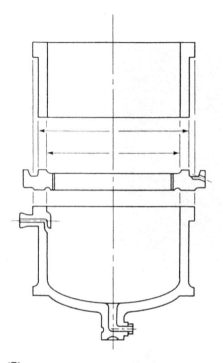

(B)

(C)

Figure 12 (B) Cutaway view showing principle of operation. (C) Top view showing perforated liquid transmission from filter. (Courtesy of Koch Engineering Company, Knight Division, 171 Kelly, Akron, OH 44309.)

(A)

(B)

Figure 13 (A) Eimco-Schriver plate-and-frame filter press. (B) Eimco-Schriver filter press showing end closure facilitator. (Courtesy of Eimco Process Equipment Co., P.O. Box 300, Salt Lake City, UT 84110.)

as to filter in one direction and wash countercurrent to the direction of filtration.

The plate-and-frame filter system can be operated at pressures of up to several hundred pounds, but it is ordinarily operated at pressures of 25–75 psi. Filter cloths of the type previously described are used, but on occasion, when it is desired to minimize contact and adhesion of the filter cake to the filter cloth, filter papers are used over the filter cloths. It was previously stated that filter cloths were used to avoid contact between the paper and the catalytic material. This factor is still a very important one, but in most cases the paper is tolerable, especially when the catalyst is to be calcined, as the presence of the paper is harmless during the calcining operation and any residual ash is also harmless in the finished catalyst. A further advantage to the use of both filter paper and cloth is that the paper prevents the filter cloth from becoming blinded with finely divided solids.

The next most frequently used type of filter is the rotary vacuum filter. Figures 14A–C show the general features of the rotary vacuum filter. This type of filter is particularly adaptable for easily filtered slurries, that is, slurries comprising a high solids content and relatively large particle sizes. Rotary filters of the size used in semiworks are usually 12 in. wide and 36 in. in diameter.

The filter medium on the drum is selected from any of the previously mentioned types. The filter drum can be operated at variable speeds so that the quantity of slurry filtered and the thickness of filter cake derived (¼–½ in.) are fairly great. As the drum rotates, there is a washing section in which sprays of water are directed against the filter cake to wash it. Following the wash there is a drying section in which air is sucked through the filter cake to remove most of the water. Thereafter, the filter cake is scraped or blown to remove it from the drum. The scraping knife can be set such that not all the filter cake is removed, leaving a residual layer on the filter medium so that the next slurry filters not only through the filter medium but also through part of the remaining filter cake. This

(A)

(B)

(C)

Figure 14 Eimco rotary filter. (A) Overall view. (B) Close-up of scraping mechanism for removal of filter cake. Filter cake may be either conveyed to dryer or reslurried in wash water for further removal of soluble salts. (C) Additional view of scraping and reslurrying action. (Courtesy of Eimco Process Equipment Co., P.O. Box 300, Salt Lake City, UT 84110.)

type of filter lends itself well to continuous operation and is applicable to commercial manufacture.

The rotary filter can be modified to what is identified as a belt-type filter, with the filter cloth actually being removed from the drum following the air suction in which the major liquid is removed. The purpose of the filter belt action is to carry the filter cake over a supplementary portion of the device so that the air can be blown through it or a spray can be applied to wash it from the filter cloth. Alternatively, the filter cake can be exposed to heating and partial drying by blown

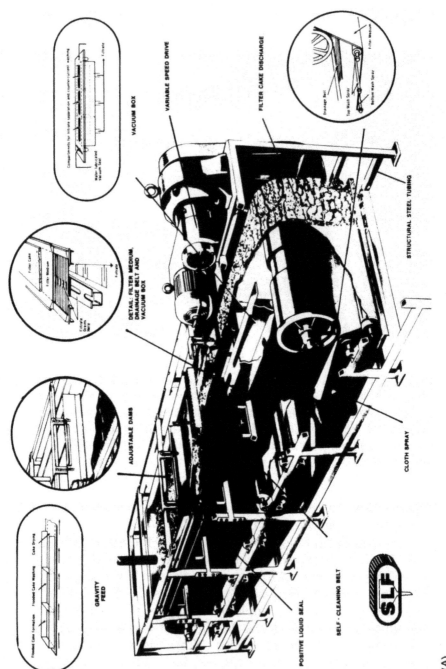

VACUUM BOX

VARIABLE SPEED DRIVE

FILTER CAKE DISCHARGE

STRUCTURAL STEEL TUBING

DETAIL: FILTER MEDIUM, DRAINAGE BELT AND VACUUM BOX

ADJUSTABLE DAMS

CLOTH SPRAY

GRAVITY FEED

POSITIVE LIQUID SEAL

SELF - CLEANING BELT

SLF

(A)

38

(B)

(C)

Figure 15 (A) Straight line filter continuous vacuum filtration and washing. (B) Filter with some enclosing panels removed. (C) Semi-works model with panels removed. (Courtesy of Straight Line Filters, Inc., P.O. Box 1911, Wilmington, DE 19899.)

air or to radiation by infrared lamps if that is preferable. Supplementary washing can also be a part of the belt-type system.

The filter cake may be discharged by passing the belt over a roll, which may be also equipped to blow the filter cake away, or a knife may be pressed against the filter cloth to remove the filter cake.

A similar type of mechanism is the line filter (built by Straight Line Filter Co., Wilmington, DE), which is very similar to the rotary filter except that instead of passing over a drum, the filter cloth moves along in a horizontal direction; this is depicted in Fig. 15. In this type of filter, a pool of slurry rests on the filter medium as it moves along in a horizontal direction. The slurry is separated by a dam from the wash water, which is separated by another dam from the drying section. The filter medium passes along over this flat table supported by a sufficiently rigid screen and grid, and a vacuum system is immediately below it.

The straight-line filter has the advantage of making it possible to filter within an easily adjusted distance by locating the dam at a greater or shorter distance from the initiation of the filtration operation. The washing and drying sections can be varied in the same way, by rearranging the dams. The rate of movement of the filter medium can be adjusted to the filtration characteristics of the slurry being processed.

A less frequently used type of filter is the centrifuge, which is shown in Fig. 16. Filtration of this type is generally possible only when the solids are grainy and easily separated from the liquid. Washing can be performed by stopping the slurry and adding wash water. The filtration, washing, and discharge can all be automated. The removal of the filter cake from the bowl can be accomplished by a scoop that scrapes the cake and removes it from the filter medium surface. An operator can also remove the cake manually.

Another filtration method uses the belt filter, in which the belt is removed from the typical cylinder so that the slurry is drawn by vacuum through the filter medium, which then passes over horizontal facilities that may provide additional washing

(A)

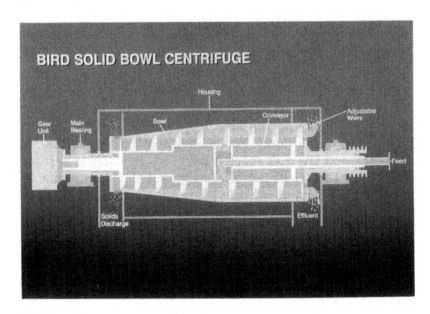

(B)

Figure 16 (A) Bird continuous centrifuge. (B) Cutaway view of this
equipment showing operating scheme. (Courtesy of Bird Machine
Co., 100 Neponset St., South Walpole, MA 02071.)

and dewatering areas. The filter cake is discharged as the belt returns to the cylinder. There are three advantages to using this type of filter:

1. The size of the particle that is filterable is smaller because the belt can be slowed down, thus permitting formation of a thicker cake and a longer time for transfer of liquid through filter cake and medium.
2. Washing can be effected much more satisfactorily and uniformly over the filter cake and medium.
3. The filter cake can be washed off the filter medium and reslurried and refiltered if it is desired to perform this type of operation to achieve a very efficient washing.

Another type of filter is the disk filter, which, in its simplest form, is a vertical disk covered on both sides with a filter medium that dips into the slurry to a depth of approximately 30% of its diameter. The disk rotates, and as it rotates through the slurry it picks up the filter cake, which then emerges at the top level of the slurry. As the disk continues to rotate, the filter cake and the disk reach a point where water is sprayed against the surface of the filter cake and washing is thus effected. An air suction section follows the wash and tends to dry the filter cake by removing excess water. The filter cake then reaches a portion of the cycle where it is scraped from the disk and drops to a conveyer or drying tray for further processing. As many as 20–50 disks can be put on a central shaft to produce an equal number of filter cakes simultaneously.

Disk filters are frequently used in the mining industry, where the slurry comprises a finely pulverized rock. As a consequence it is obvious that the catalyst that is processed through this type of device must be an easily filtered, rather grainy type of material.

5

Drying: Containers, Trays, and Other Drying Auxiliaries

Preliminary to a consideration of the drying facilities, one must consider the containers in which the filter cake or other material to be dried is to be placed. In the laboratory one usually employs an evaporating dish with a simple glazed surface. If, however, the catalyst is to go into the calcining operation immediately after the drying and not placed into another type of container, it is highly desirable to have an unglazed pure alumina or silica container. Both of these are quite expensive but are well worth the cost as they avoid the adverse effects of a glaze reacting with the catalytic material.

If the catalyst being dried does not have poor corrosion characteristics, then the containers can be fabricated from stainless steel, nichrome, or inconel. Nichrome or inconel is employed when the calcining operation is carried out at temperatures in excess of 600°C. When the size of the operation is increased to pilot plant or commercial operations, then it is essential that metallic trays be used.

In many cases the trays that are employed are enameled steel if the glaze itself does not present a problem. Enameled steel is, however, subject to easy fracture and crazing in the hands of careless operators. It also has the problem of any glaze in that at higher temperatures it either softens or will react with the catalytic material, generally in a harmful way. It is much better to employ unglazed alloy trays if it is at all possible.

Furthermore, if it is desirable to have a continuous belt dryer, then a flexible metal can be used as the belt material. It may be possible to use Teflon-coated metal or even Teflon itself, but this requires special consideration and design. The trays are ordinarily loaded to a depth of not more than 1 in., and it is even preferable that the depth be limited to less than 1 in. It is also desirable that the trays be placed in racks containing from five to 30 trays, one above the other. These should be so situated that the free space between the top of one tray and the bottom of the tray above is equal to at least twice the depth of a tray. This will permit the air or other drying medium to pass with a minimum of restriction above the catalyst as it is drying. The location of the racks of trays in the drying equipment is discussed later with the drying facilities themselves.

Catalysts can also frequently be dried by so-called spray drying or flash drying. These drying procedures are discussed later in the chapter and in Chapter 10.

DRYING EQUIPMENT

The most frequently used drying equipment in the laboratory is the simple laboratory oven. Ovens can be purchased with multiple shelf possibilities and with temperature variation and control between 75°C and several hundred degrees Celsius. They can be either vacuum ovens (Fig. 17) or controlled atmosphere ovens. All types have a place in the catalyst development laboratory.

The simplest oven is an enclosed space with a electric heater at the bottom. The gases, usually air, in the oven are heated

Figure 17   Pennwalt–F.J. Stokes vacuum oven. (Courtesy of Stokes Vacuum, Inc., 5500 Tabor Road, Philadelphia, PA 19120.)

from below, and convection allows these heated gases to pass up over or through the catalyst to remove moisture or other volatile materials. The gases are exhausted through openings at the top of the oven or around the doors. Usually there are both temperature indicators and temperature controllers in the oven to permit the selection and maintenance of temperatures for a given type of catalyst. In cases in which simple drying alone is needed, the temperature selected may not exceed 110–125°C, but, on the other hand, where there is combined water such as in the case of gels, the temperature desired may be as high as 200–250°C.

A more sophisticated oven is one in which previously selected gases are admitted at one portion and exhausted at another (Fig. 26). The most commonly used gases are carbon dioxide, nitrogen, or some of the rare gases such as helium or argon. In other cases when the drying is carried out under vacuum, the temperature can be regulated to a subfreezing level, in which case the catalyst is freeze-dried. Vacuum-drying conditions are generally selected when the catalyst is temperature-sensitive and must be processed at a temperature below 100°C.

Freeze-drying is sometimes practiced when the catalyst is at least partially in the form of a gel. The gel will coalesce to form hard, dense granules having very small pores, whereas if the water is removed while frozen, these particles cannot coalesce and the powder resulting from the drying operation is a finely divided product free of the shrinkage that causes the development of the small pores. Pore size can be controlled by altering freeze-dry conditions and by using liquids other than water, such as alcohols, glycols, esters, or aqueous solutions of them.

SEMIWORKS DRYING FACILITIES

The simplest drying facilities for semiworks are very similar to the simplest equipment described for laboratory facilities, which consists of a suitably sized cubicle with door and heating element either at the top, with a fan circulating the hot gases down over the catalyst, or at the bottom or sides of the enclosed cubicle (Fig. 17). In the semiworks, as in the plant-scale equipment, the cubicle can be heated electrically, by steam tubes, by means of hot air from a direct flame combustion heater, or by heat exchange of air from the products of combustion of the flame.

In either the semiworks or plant systems in which products of combustion are directly contacted with the catalyst, there is a possibility that incompletely burned hydrocarbons, specifically carbon monoxide, will contact the drying catalyst. This can have a very harmful effect on the catalyst, and even carbon dioxide is on occasion a factor that must be considered and avoided. To avoid direct contact between the products of combustion and the catalysts, a heat exchanger is frequently used, with air circulated to the heat exchanger and the products of combustion confined to the exterior part of the tubes of the exchanger. This is ordinarily quite satisfactory, but there are occasions when the heat exchanger will corrode badly enough to allow the gases to be transmitted into the circulating gases in the drying unit, with the result that the products of combustion come into contact with the catalyst, bringing about the harmful effect the use of the heat exchanger was intended to avoid.

One of the difficulties in using ovens in which the gas is rapidly circulated through a heat exchanger or electric heater or steam coils is that the air that is being circulated may pick up particles of catalysts, which then contaminate the circulating system. There is the possibility that such contamination will eventually migrate back into the drying chamber after a different catalyst has been placed in the unit. If one is processing an alcohol synthesis catalyst, for example, and the circulating air contains particles of a previously prepared methanation catalyst, it is obvious that the alcohol catalyst will be very severely and adversely affected by the cross-contamination. This cross-contamination can be avoided, of course, by isolating certain pieces of equipment for use in only certain types of catalyst manufacture, or, if this is impractical, then means should be set up whereby all of the circulating gas system in the dryer can be washed or vacuumed out very carefully.

It is very discouraging to be drying a white catalyst and find a film of brown or black oxides of a different type on the surface of the catalyst. Not only is it discouraging with respect to the white catalyst, but also it indicates that if a catalyst of a dark color were being processed, it is possible that a different catalyst of a dark color might be contaminating the surface but would not be evident because of its similar color. The factor of cross-contamination in drying and calcining equipment simply cannot be overemphasized.

We have described the simplest forms of drying facilities and have discussed the problems of cross-contamination and means of avoiding this contamination. Now we proceed to other types of dryers that differ substantially from the compartmentalized drying systems.

The first is the belt dryer, which is useful for continuous operations and is frequently used in processes in which the catalyst at one stage is extruded. The extrudate is dropped immediately onto a moving belt, the motion of which must keep the extruded wet pieces from building up into piles or clusters that merge together to form a lump. When such a cluster is dried, it remains an unmanageable coalesced lump; the lump can be broken apart, but usually in such a way that dust and unacceptably small particles are formed. Therefore, the belt must be moving at a rate well synchronized with the rate of discharge of the extruding device. The belt will then have just one layer of extruded pellets that proceed through a drying zone in which the moisture is removed and the particles become discrete and noncoalescing entities. The oven is generally heated by resistance elements or lamps radiating directly down upon the moving belt and catalyst particles or by steam heat or circulating air that passes directly over the catalyst as it is moved through the dryer on a belt.

It could be said at this point that in conjunction with the drying segment of the over, there can be a downstream portion of the oven that is heated to a much higher temperature and in which calcining can be carried out. Small calcining equipment of this type is available, but at considerable expense because it is custom-made, usually to meet the demands of a particular process. The pilot equipment may consist of a linear chamber accommodating an endless moving belt 6–10 in. in width and 20–30 ft. in length in the heated zone. Plant equipment, as later described, is substantially larger, but it also is ordinarily made to custom requirements.

TUNNEL DRYERS

Tunnel dryers (Fig. 18) are also continuous facilities and usually are employed for extremely high temperature processing. They are usually used in the ceramics industry and as a consequence are frequently used for catalyst supports such as rings, rods, and cylinders. The tunnel dryer is generally designed and built in such a way that a bucket conveyor or a train of individual cars on a track will pass through a heated or high-temperature fired tunnel. These are used at temperatures ranging from drying temperatures of 100–200°C to calcining or firing temperatures, which may be in the range of 800–1700°C. When these very high temperatures are required, then the use

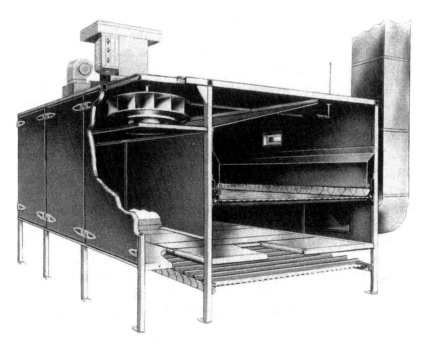

Figure 18 Proctor & Schwartz tunnel dryer. (Courtesy of Proctor & Schwartz, Inc., 7th and Tabor Road, Philadelphia, PA 19120.)

use of metals, of course, is precluded and everything must be constructed of ceramics. As was indicated earlier, the tunnel dryer is most useful in those cases in which a very high temperature system is required and usually for catalyst support rather than for the catalysts themselves. This is not, however, to preclude the processing of impregnated ceramic materials.

DRUM DRYERS (SPECIALLY DESIGNED ROTARY OR BLANKET FILTERS)

A drum dryer is used when the catalytic or support material is in the form of a mudlike composition. A puddle of the mud is situated and kept agitated directly below a rotating drum. As the drum rotates and dips into the mud, a layer of dried product forms on the outer surface of the drum. The drum is ordinarily heated by superheated water, Dowtherm, or some other high-temperature liquid. Steam can also be used, but this is usually subject to some difficulties, such as the pounding and crackling of the steam as it condenses in the drum itself. As the drum rotates, the mud is dried to the point where it can be scraped off as a dry powder or cake. This is usually just before the surface of the drum is reimmersed in the mud in the pan below the drum.

The drum dryer is frequently used in cases in which a dry- and filtering operation can be combined, eliminating filtration. Instead of filtration, a thickening operation is sometimes performed using a specially designed rotary filter (Fig. 19) or a thickener. It must be recognized that as the catalytic material is scraped from the surface of the drum, it is apt to be contaminated by the drum itself. Proper precautions to avoid this are necessary and usually are accomplished by ensuring that the materials employed for construction of the drum are tolerable in the catalytic materials. It should also be noted that if there are residual salts dissolved in the liquid phase of the slurry, those salts will also dry into the product powder and may affect the acceptability of the finished catalyst.

(A)

(B)

Figure 19 (A) Sweetland filter. (B) Front view. (Courtesy of Dorr-Oliver Incorporated, 612 Wheeler's Farm Road, Milford, CT 06460.)

On the positive side of the ledger is the likelihood that the drying on the drum can be accompanied by compression rolling at one stage as the drum rotates; this compression rolling can also eliminate the need for a densification step for the powder. Thus, it can be seen that the drum dryer can prevent the need for filtration as well as densification, but it should be recognized that it is relatively crude processing that may not be adaptable to the most sensitive and sophisticated catalysts.

ROTARY DRYERS

A rotary dryer is adaptable only to semiworks and commercial catalyst manufacturers. The semiworks rotary dryer may be as small as 18 in. in diameter and no more than 8–12 ft long, whereas the commercial rotary dryer may be as large as 10 ft in diameter and 90 or more feet in length. The largest rotary dryers, however, may be much too large for any catalyst manufacturer. These large dryers are used primarily in the cement or agricultural lime industry.

The basic principle of the rotary dryer is that the material to be dried is fed in at the upstream end of the cylinder and discharged at the downstream end. The dryer can be heated either by combustion gases passing cocurrently or countercurrently with the material being dried, or there can be steam or electric coils along the outer perimeter of the dryer.

Problems encountered in the use of the rotary dryer include the caking of wet or semidried catalysts at or near the point of introduction of the wet cake. This is usually handled by having an automatic mechanical hammer periodically strike the outer circumference of the dryer and shock the dried adhering cake loose from the walls. The dryer thus can be freed of the adhering material, but because of its inherent caking tendencies and because of the fact that it is broken from the wall as a lump, it may remain as a lump as it passes through the dryer. The center of such lumps may remain wet so the drying process is incomplete.

It is obvious that one of the most severe difficulties met with in using equipment of this type has to do with the cleaning following the processing of one type of catalyst before the initiation of the processing of another type. Consequently, it is highly desirable, if economically justifiable, that specific pieces of equipment be set aside for specific catalysts. Rotary dryers are particularly applicable for large-volume products such as alumina hydrate or molecular sieves or for the iron chrome type for either high-temperature water gas shift or the dehydrogenation of ethylbenzene to styrene. It is obvious after this description of the types of materials processable that the preparation of a white material such as alumina hydrate following the processing of an iron oxide type of catalyst would require meticulous cleansing of the equipment. This observation becomes axiomatic in view of the fact that iron contamination of activated alumina is one of the most severe problems encountered when the latter is used in many operations.

FLASH DRYER

It is almost impossible to describe the flash dryer without reference to a diagram or a picture. Consequently, reference is made to Fig. 20, which depicts a comparatively easily comprehensible representation of the principle and operation of a flash dryer. The flash dryer is used primarily in the processing of large-volume catalysts such as those used in hydrotreating and cracking and the clays frequently used as binders in these types of catalysts.

A further application for the flash dryer is in the drying of those unique catalysts that must be processed quickly from a slurry or liquid phase to a powder phase so as to minimize the time at which they are at drying temperature in a moist condition. Some catalysts are very sensitive to the action of steam or water during their exposure to temperatures near 100°C, and as a consequence the flash dryer is a highly satisfactory processing facility for them. The flash dryer is sub-

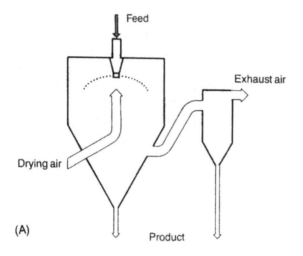

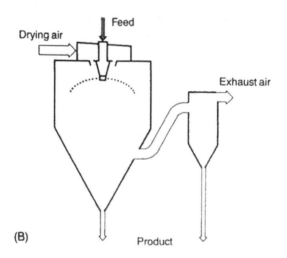

Figure 20 (A) Spray dryer schematic diagram showing rotary atomizer and central cooled air disperser. (B) Spray dryer schematic diagram showing rotary atomizer and roof air disperser.

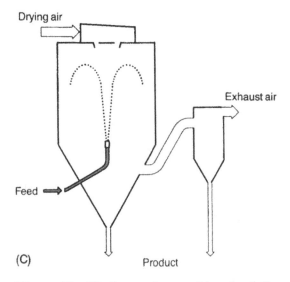

Drying air

Exhaust air

Feed

(C)   Product

Figure 20 (C) Spray dryer with mixed flow and nozzle atomizer. Schematic diagram. (Courtesy of Niro Evaporators, Inc., 9164 Ramsay Road, Columbia, MD 21045.)

stantially different from the spray dryer, which has a large number of applications and is the next type to be described.

SPRAY DRYER

Spray drying is broadly used for the processing of fluidized catalysts as well as for some small-particle catalysts or supports such as microspheroidal silica and alumina. This process is best described by reference to the diagram and pictures in Fig. 21. The operating conditions for the spray dryer are very broad with respect to temperature, gas used, height and (especially) diameter of the spray dryer, the size of microspheres, and the narrowness of mesh size of the microspheres that are produced.

(A)

(B)

Figure 21 (A) Industrial installation of spray dryer. (B) Spray dryer section of industrial installation of spray dryer.

(C)

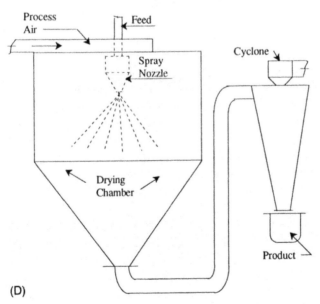

(D)

Figure 21 (C) Collector section of installation of spray dryer. (D) Schematic diagram illustrating spray nozzle type spray dryer.

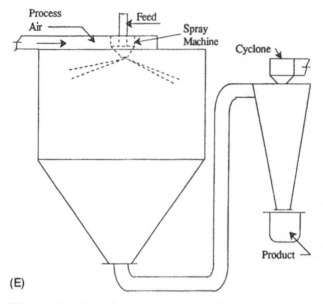

Figure 21 (E) Schematic diagram illustrating rotary spray machine type of spray dryer. (Courtesy of Niro Evaporator, Inc., 9164 Ramsay Road, Columbia, MD 21045.)

The process is particularly adaptable to gels, which on drying tend to shrink and harden; in fact, many catalysts are intentionally made into a gel so they can be processed in the spray dryer. When in normal processing of the catalyst it is impossible to derive a gel, a gel state is induced by including in the composition of the catalyst a colloidal silica or a silica or alumina gel or other gels that tend to shrink and harden as they dry.

Spray dryers can be purchased in sizes as small as about 4 ft in diameter, which are specifically designed for laboratory use, but also may be as large as 12 or 15 ft or even larger in diameter and at least this same height. The operation of the spray dryer is dependent upon the type of material being processed, and as a consequence if one wishes to operate continuously, it is essential that the feed to the unit be uniform as to moisture content, solids type, and solid characteristics. The slot size and speed of rotation of the spray heads must also be constant and selected after careful expert analysis, usually after laboratory studies of the materials and equipment in question. As one might guess, the spray head is especially designed for specific types of operation, and the speed of rotation is also selected for given characteristics of the feed slurry. It is apparent that if an abrasive material is being processed, the openings on the spray head may erode to the point where the character of the spray droplets changes, which in turn alters the character of the product. The spray dryer can probably best be described as one of the processing steps in catalyst manufacture that responds mostly to art rather than to science.

A person well skilled in the spray dryer's operation can do much to ensure uniformity of product and operation.

6

Calcining

In laboratory preparation of catalysts there are two general types of equipment used for the calcining operation. The first is the simple muffle furnace shown in Fig. 22, which consists of a housing that covers insulation that in turn covers the small cubicle where the heating takes place. These furnaces are built by several manufacturers and have a very common and typical appearance. The interior is usually heated by means of a nichrome coil embedded in a ceramic plate. In many cases the heating coil is not completely embedded but exposed, and in this case it is difficult to clean. Many catalysts, after calcining, are very low density powders and may be partially lifted from the container by gases evolved during calcining. This dust may migrate to the coils themselves, from which it is difficult to remove, or even worse, may fuse onto the coils, thereby destroying them. This is usually avoided in the laboratory by having a totally closed insert fabricated from stainless steel

(A)

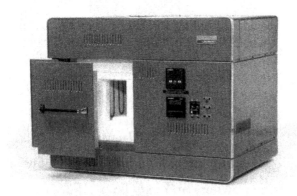

(B)

Figure 22 (A) Lindberg laboratory muffle furnace. (B) Front view with door open. (Courtesy of Lindberg – A General Signal Company, 304 Hart Street, Watertown, WI 53094.)

or other high-temperature-resistant alloy. This container iso-
lates the heating elements from the catalyst being processed.
Furthermore, the insert can be so designed that it is gastight,
which permits the processing in either a controlled atmosphere
or under elevated or reduced pressure as desired. The controlled atmosphere can, of course, be steam, oxygen, carbon dioxide, nitrogen, hydrogen, or carbon monoxide. In some cases
even ammonia or H_2S can be used when it is desired to nitride
or sulfide the catalyst. Methane or carbon monoxide permits
carbiding of the catalyst.

The variety of commercial calciners is depicted in Figs.
23 and 24.

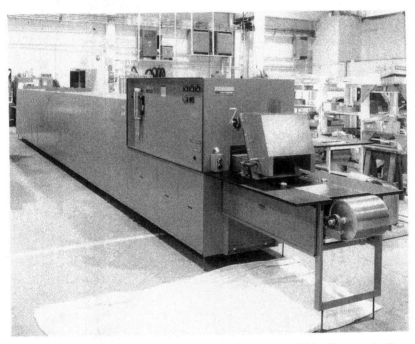

Figure 23 Lindberg tunnel calciner. (Courtesy of Lindberg – A General Signal Company, 304 Hart Street, Watertown, WI 53094.)

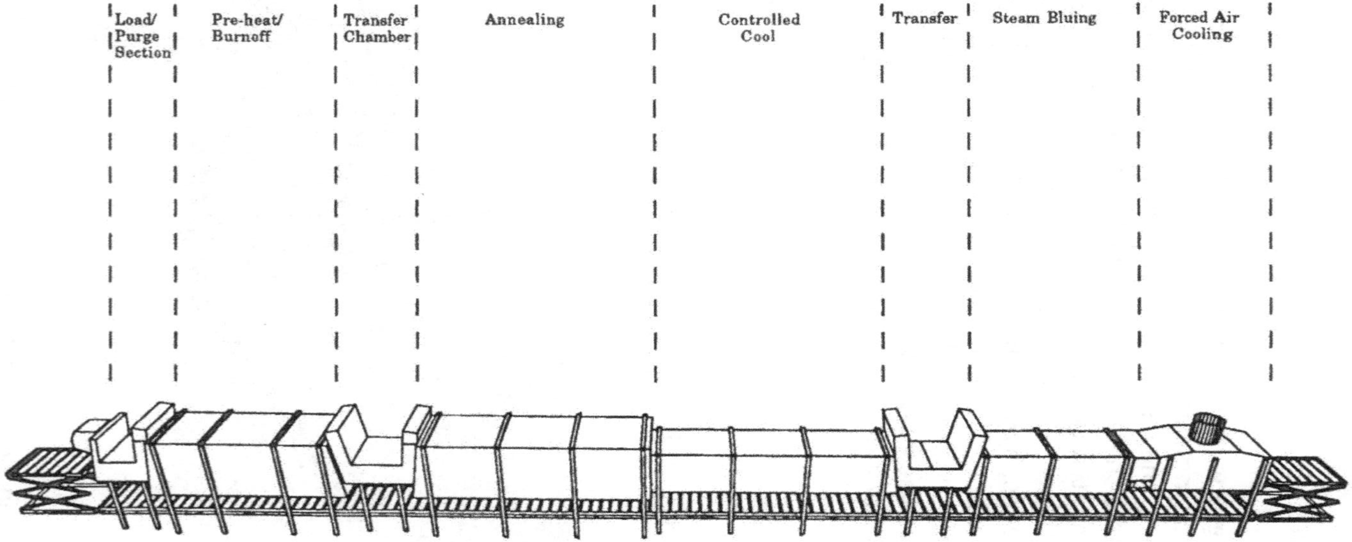

Figure 24 Roller hearth annealing furnace. Diagrammatic representation and labeling. (Courtesy of Lindberg – A General Signal Company, 304 Hart Street, Watertown, WI 53094.)

Load/Purge Section

Pre-heat/Burnoff

Transfer Chamber

Annealing

Controlled Cool

Transfer

Steam Bluing

Forced Air Cooling

Furnaces heated with a nichrome coil are, of course, limited to temperatures below 1100°C. If operating temperatures up to 1400–1500°C are desired, the heating can be accomplished by the use of platinum rhodium or other precious metal alloy as the heating element. The containers for the catalyst must be either fused silica or fused alumina to withstand temperatures in this range.

It is also possible to heat the furnace by means of silicon carbide rods (globars), which also permit operation up to 1400–1500°C. This may be desirable when processing catalysts such as supported precious metals that are to be used as substitutes for precious metal gauze in ammonia oxidation or HCN synthesis.

Another type of furnace frequently used in the laboratory is the so-called split tube furnace, which is shown in Fig. 25. This furnace has controls and heating elements similar to those previously described as being embedded in ceramic blocks or plates. In the case of the tube furnace, the nichrome wires are coiled in grooves in the walls of ceramic cylinders of a size necessary to derive the tube opening of specific furnace length and diameter. One specific advantage of the split tube furnace is that the halves are hinged and can be opened during the heating process and while they are hot. If a transparent Vicor or silica tube is used, color and texture changes can be observed that may be valuable to note. Furthermore, the tube can be removed from the furnace under the controlled atmosphere. It can also, of course, be inserted into the furnace under similar conditions.

The tubes are generally fabricated from Vicor, which is operable up to approximately 700°C, or quartz or alumina, which can be operated up to the limit of the nichrome wires. Usually the catalyst is placed in a boat fabricated of either silica or alumina, and the boat is inserted into the tube in such a position that it is essentially midway in the furnace itself. Both ends of the tube are closed to the atmosphere, but each end is designed in such a way that gas can enter at one end and exhaust at the other. This obviously makes possible a con-

(A)

(B)

Figure 25 (A) Lindberg split tube furnace with controls. (B) View
showing all components of the Lindberg split furnace type system.
(Courtesy of Lindberg – A General Signal Company, 304 Hart Street,
Watertown, WI 53094.)

trolled atmosphere within the tube. There is also usually a thermocouple well that extends from one end to the other and allows a temperature reading to be obtained at the position where the boat is located.

In addition, there is usually a thermocouple on the heater side of the tube to determine the difference in temperature between the interior of the tube and the furnace heating element.

This, of course, has the purpose of minimizing overheating which could occur because both the catalyst and the refractories used are comparatively good insulators. A temperature differential of substantial magnitude could develop between the heating element and the interior, causing the catalyst to overheat. An additional reason for having a thermocouple in the proximity of the catalyst as it is being heated is to enable the detection of endotherms or exotherms experienced by the catalyst. For example, when basic chromates are decomposed to chromites, there is a substantial exotherm.

It is worthwhile to determine the extent of the exotherm and ensure that the catalyst has not risen to a temperature that could be harmful to it. Analytical techniques used to control the quality of catalysts include DTA (differential thermal analysis), which scientifically reveals what may be a series of endotherms and/or exotherms that in turn reveal phase changes or decompositions important to catalyst quality and uniformity.

TUNNEL CALCINERS

In the description of drying equipment in Chapter 5, the tunnel dryer was described, and at that time it was indicated that in addition to the drying section of a tunnel dryer there could also be a calcining section in which the temperature was substantially higher. The tunnel dryer ordinarily is designed in such a way (Fig. 26) that the material to be calcined is placed either in buckets or in small carts that are linked together by chains or other mechanisms and are loaded at the inlet end of the dryer-calciner and discharged automatically at the exit

Figure 26 A diagrammatic presentation of Lindberg roller hearth controlled atmosphere continuous annealing furnace. (Courtesy of Lindberg – A General Signal Company, 304 Hart Street, Watertown, WI 53094.)

end. The buckets or carts are recycled to the entrance end via a continuous chain or sprocket device.

This device permits operating at probably the highest temperature of any of the calcining equipment because the buckets or other containers can be fabricated of ceramics and passed along ceramic rails or a ceramic surface, thus eliminating the exposure of (the lower melting point) metals to any of the heated areas.

BELT CALCINER

The belt calciner was described in Chapter 5 as a belt dryer but reference should be made to Figure 26 for a visualization as implied in the preceding section on the tunnel calciner. The belt type of calciner has a temperature range substantially below that of the tunnel type of calciner. This is only a minor limitation, because most catalysts are processed at temperatures that are acceptable to the high-temperature alloys such as a inconel or nichrome. Some of these alloys have been improved in recent years by the addition of a small amount of cobalt.

The continuous belt can be fabricated from many different kinds of metals and in many different designs. In those cases in which a catalyst that is in the form of fairly large granules is being calcined, it is desirable to use a chain-like belt that permits gases to circulate readily through the belt and the catalyst itself. This is particularly useful, as will be subsequently discussed, when the catalyst has been made specifically into pelleted, granular, or extruded form. In most of these cases there is a small amount of lubricant such as graphite or an organic compound such as stearyl alcohol or stearic acid present, and these may burn out during calcining, with the result that excessive temperatures are reached. In such a case, the atmosphere surrounding the catalyst is controlled to limit the presence of oxygen and hence oxidation. In other cases the temperature rise is "programmed," the temperature being slowly raised so as to slowly oxidize the carbonaceous or combustible material present.

The only suitable method to employ when the catalyst itself contains a reducible oxide such as manganese or copper oxide is the carefully controlled (programmed) temperature rise. When these oxygen sources are present, they can react with the oxidizable component of the catalyst in a drastically harmful manner. The calcining of formed pellets will be discussed more fully subsequently.

ROTARY KILN AS CALCINER

The rotary kiln was described previously under drying equipment (Chapter 5). Reference should be made to this description for a broader understanding of the discussion that follows. The rotary kiln is a relatively crude method of calcining and is applicable only to those cases in which a rather wide variation in temperature and atmosphere is tolerable. Furthermore, the gases that are used for heating are in essentially every case in direct contact with the material being calcined or heat-treated. As previously stated, such equipment is useful only for the crudest of catalysts or catalyst supports, such as alpha-alumina, periclase, mullite, diaspore, or bauxite.

7
Rewashing and Ion Exchange

Metal catalysts, when precipitated, have a tendency to occlude foreign ions such as sodium or other alkali or an anionic component such as sulfate or chloride. Usually, volatile components such as nitrate or carbonate are not considered to be harmful occluded ions. However, if the precipitate is a carbonate, hydroxide, or gel-like type, these ions (SO_4, Cl, etc.) will be so firmly occluded that washing simply will not remove them adequately. An alternative to simple washing of the fresh precipitate is to filter the catalyst and dry and calcine it. The calcining frees the occluded ions, which then ordinarily can be removed by washing with water or even more completely by ionic exchange with a salt such as ammonium carbonate. The ammonium ions will exchange with cations, and the carbonates with anions. The occluded ions can now be washed away and replaced in the catalyst with the volatile NH_{4+} and CO_3^2. On drying the catalyst after this exchange, the now-present volatile ions are volatilized away.

In the laboratory this ion exchange is performed by agitating in a beaker. The ion exchange can be effected at essentially any temperature that does not promote the decomposition of the ammonium carbonate. Because ammonium carbonate decomposes at temperatures in excess of 75°C, it is wise to conduct the exchange at 25–50°C. It is also desirable to have the catalyst finely divided to facilitate ionic exchange as it is slurried. Lumps of catalyst can usually be broken up by the agitation of the slurry itself, but when agglomerated particles are so hard that this is not effective, it is desirable to pulverize the catalyst prior to the ion exchange. Pulverization can be effected in a rotary disk or ball mill or even by extremely rapid agitation in the rewashing facilities.

It will be recalled that warning was given previously to the effect that it would be harmful in some cases to have excessively rapid agitation in the precipitation step, and it was recommended at that time that a paddle-type agitator with slow peripheral speed be used in the precipitation environment. By contrast, when the catalyst is ion-exchanged in an aqueous or other solution, it is not essential at this time to avoid particle disintegration as was the case with the initial precipitate, and as a consequence rapid agitation such as that effected by the propeller-type blades would be quite satisfactory or even preferred. There are other types of agitators such as turbine or fan (multiblade) types that also would be acceptable.

REWASHING AND ION EXCHANGE IN SEMIWORKS AND COMMERCIAL SCALE

The principles, of course, are the same for catalysts on the semiworks and commercial scale as for those on the laboratory scale. Many people viewing a catalyst plant for the first time say it is nothing but an overgrown chemical laboratory with tanks instead of beakers, agitators instead of stirrers, presses or rotary filtration instead of filters, and so on. The preferred agitator for commercial manufacture was previously identified as one of the paddle type; the propeller and turbine types were said to cause excessive shearing. By contrast, the shearing action that has an adverse effect on the original precipitate has the effect one wants to produce when breaking down lumps to facilitate ion exchange. As a result, the propeller or turbine types are both satisfactory designs for the reslurrying operation. The shearing effect, if excessive, can be altered by changing the speed of rotation. If a given tank were to be used for various types of precipitates, with one type of catalyst requiring one speed and another catalyst requiring a different speed, then this could be achieved by using a variable-speed drive.

The commercial and semiworks scales are not described individually here because the commercial equipment is essentially a larger version of the semiworks facilities.

8

Densification of Calcined and Ion-Exchanged Catalytic Material

It has previously been stated that if a catalyst is to function in fixed-bed operations, either gas or liquid phase, it must be converted to relatively large particulate form by a granulation, extrusion, or pelleting process. The pelleting operation is performed in what is described in the pharmaceutical industry as a pill-making machine. For satisfactory operation of the pilling process, the powder feed must have definite characteristics (specifications) of density, moisture content, particle size distribution, and compaction character and must contain die and punch lubricants. The catalysts after precipitation and calcining are, in essentially every case, of too low a density to be processed in a pilling machine. However, catalysts that have been through a rewashing operation (ion exchange) may in some cases be sufficiently dense for pilling without further densification. A simple preparation of the powder for pilling and testing will readily establish this.

Tests of compaction of powders indicate the desirability of the powder being made up of a mixed particle size so that what passes an 8 mesh screen, for example, may very well have a major fraction of that powder as a 100–200 mesh fraction. This fraction can be compared to the sand in a sand, gravel, and concrete mix. The sand adds to the agglomeration and strengthening of the final concrete structure just as the powder fitting between the larger granules of catalyst tends to agglomerate and strengthen the larger particles in the final compression and pilling operation.

It is generally true that the optimum screen size distribution of the powder mix is different for different catalyst powders. Some catalyst powders compress and become strong with almost any screen size distribution. These catalysts generally are those that contain copper oxide, zinc oxide, aluminum oxide, or magnesium oxide. These oxides pill with little effort, whereas copper chromite, nickel chromite, and some molybdates are at the extreme opposite end and require the most meticulous care when preparing the powder for pelleting. Insufficient or excess moisture, low density, and excessively fine powder will cause severe pilling problems. Sometimes pilling aids are beneficial.

PILLING AIDS

In order for a catalyst to pill satisfactorily, a primary requirement is that a lubricant, previously identified as graphite, Sterotex (a trade name for stearin), or other aids yet to be described, be present. It is not the practice ordinarily to add a lubricant during the densification stage. However, with some types of powders, for example, copper chromite, it is essential that a pilling aid be added.

These pilling aids frequently are put into the catalyst at the time the catalyst is densified, an operation that will be soon described. Those materials that have found to be satis-

factory are sodium silicate, potassium silicate, calcium aluminate, sodium aluminate, magnesium hydroxide, magnesium
oxide, and some chromates such as magnesium or calcium
chromate.

DENSIFICATION OF POWDER FOR
PELLETING, EXTRUSION, OR GRANULATION

Densification on the laboratory scale can be brought about in
a number of different ways. If there is nothing more suitable
available, a large mortar and pestle can be used. When the
mortar and pestle is used, the powder in question is put into
the mortar along with a small amount of water. This is ground
with the pestle until it becomes a homogeneous dense paste
resembling putty or a thick clay mud. When the paste is removed from the mortar and dried, it should have become much
harder and denser so that in crushing and passing it through
an 8 mesh screen, a dense granular powdery mix is derived
that is suitable for pilling.

On occasion, the powder after this operation is not sufficiently dense, and a simple expedient in that case is to put the
paste from the mortar into the center of a heavy duck or twill
filter cloth. The filter cloth is folded over and over against itself so as to enclose the paste in a cloth envelope. This envelope is usually folded in such a way as to derive a square approximately 4 in. on the side. This square is next sandwiched
between two ¼-in. stainless steel plates, also 4 in. on the side.
This sandwich is now placed in a large mechanical vise and
screwed tight so that the water is squeezed from the paste
and the paste is converted to a relatively dry, heavily compressed sheet. This sheet is then removed by unfolding the
filter cloth envelope, and it is dried in a suitable container such
as an evaporating dish. This material generally dries to a very
hard and suitably densified cake for later pilling. This process
can be repeated to derive sufficient material for a pilling test.

In some cases a promoter or a pilling aid should be added, and if this is the case, then instead of adding just water to the powder in the mortar as earlier described, one first dissolves in the water a promoter or pilling aid. As an example, potassium carbonate may be added if the catalyst is to find use in the synthesis of higher alcohol from CO and H_2. Potassium carbonate is a promoter for the derivation of higher alcohols, but it is also a pilling aid. In addition to or instead of the potassium carbonate, potassium dichromate or potassium permanganate can be used, both of which are higher alcohol promoters as well as pilling aids.

SEMIWORKS DENSIFICATION

There are many ways to achieve densification on the semiworks scale. The preferred one is to use a scaled-down kneader of the type shown in Fig. 27. There are several manufacturers of small kneaders; the one pictured is manufactured by Teledyne-Readco and is identified as a Readco mixer.

In this operation the low-density catalyst powder is placed in the bowl of the kneader, and then the blades are rotated while small portions of water are added to cause the powder to be converted to a paste. Care must be exercised to avoid excess water, or the paste will be overwet and as a consequence, on drying, will not be sufficiently dense for it to produce a powder having good pilling characteristics. The general practice is to put into the bowl only about two-thirds of the weight of the powder to be kneaded, and then, as this becomes excessively wet, which is the typical problem, add the remaining one-third of the dry powder slowly to the kneader over a period of 5–10 minutes, with the result that the paste dries to the point where effective kneading and densification take place.

In the early stages of the kneading operation, the powder often seems to be too dry and additional water is added, but with even a small amount of additional water, the paste becomes excessively wet. Additional water should be added only

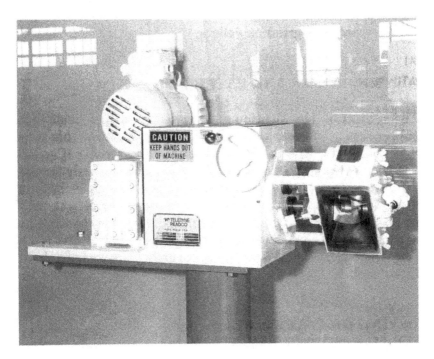

Figure 27 Teledyne Readco laboratory kneader in discharge position. (Courtesy of Teledyne Readco, 901 South Richland Avenue, P.O. Box 15552, York, PA 17405-0552.)

after 5 or 10 minutes of kneading and it is evident that a dense paste will not otherwise be attained.

Note: The quantity of water required is generally in the range of 30–50% of the weight of the powder. That is, if 1000 g of powder is being kneaded, between 300 and 500 g (300–500 mL) of water should be used for the densification. Successive lots of the same powder should require the same amount of water. If more or less is required, one should suspect some changes in precipitation, calcining, or storage conditions. Most catalytic powders are highly effective desiccants.

The catalyst can also be densified by slugging in the tabletting machine, and this will be subsequently described. The catalyst can also be densified by being passed between two compression and grinding rollers.

DENSIFICATION AND EXTRUSION

The subject of densification and extrustion will be dealt with subsequently in greater detail in Chapter 9, but at this time it is noteworthy that one can densify in an extruder (Fig. 28) and then extrude the moist, densified paste as variously shaped extrudates that can then be cut to form pellets with length and diameter essentially equal, or in some cases the length may be 50–200% greater than the diameter. It is also possible to extrude the product and cut it into pellets with lengths less than their diameter. All of this is covered more completely in Chapter 9.

DRYING OR CALCINING
FOLLOWING DENSIFICATION

If the only purpose of the drying operation following densification is the removal of water, then the temperature is usually in the range of approximately 140–160°C. Most catalysts have a strong propensity to retain water, so it is necessary to attain temperatures in excess of 110°C for this strongly adsorbed, or even chemisorbed, water to be evolved.

   If during the course of the densification some promoters are added, it may be necessary to calcine the catalyst at temperatures at which the added promoters are decomposed. This is usually below 400°C, although in exceptional cases the catalyst may require a temperature of 450–500°C. The problem that may arise as a result of this high-temperature calcining is that the subsequent pelleting or pilling of the catalyst may be adversely affected by the fact that the powder is excessively dry. The readsorption of moisture to a controlled degree (say,

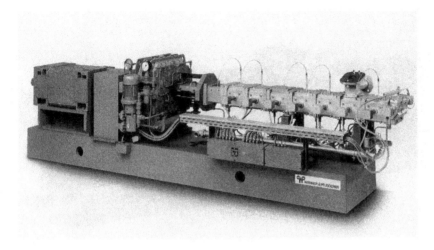

(A)

(B)

Figure 28 Werner & Pfleiderer extruder. (A) View showing mechanism with cover removed. (B) View of opposite side of machine.

(C)

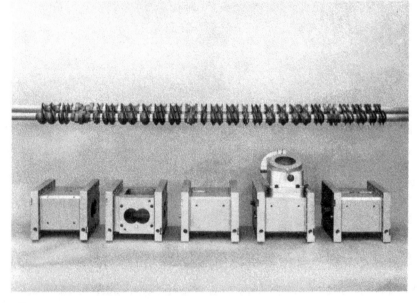

(D)

Figure 28 (C) Cutaway view showing extruder screw. (D) View showing disassembled parts – extruder screw, and sectional barrel.

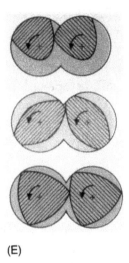

(E)

(F)

Figure 28 (E) Cross-sections of corotating screws. (F) View showing section of barrel and screws. (Courtesy of Werner & Pfleiderer Corporation, 663 E. Crescent Avenue, Ramsay, NJ 07446.)

3-6%) sometimes is quite difficult and may necessitate special steaming or humidification equipment. Ordinarily, this is not necessary, and letting the catalyst product stand at atmospheric conditions for a predetermined period is usually sufficiently effective at rehydration that the pilling operation can be performed normally.

9
Pulverization, Pilling, and Extrusion

The dried densified powder is typically crushed to 100% through a definite screen size using equipment of the type shown in Fig. 29 or a ball mill (Fig. 30), with the screen size depending upon the size of the pills to be prepared. For example, for ½ or ¾-in.-diameter pellets, powder is passed 100% through a 6 mesh screen, whereas if it must find its way into ⅜-in. dies to make ⅜-in. pills, then the powder should be passed 100% through a 10 or 12 mesh screen. Finer powder is generally processed to extrudates.

MIXING WITH PILLING LUBRICANT

After the powder has been sized to be suitable for feeding to a pilling machine, it is mixed with lubricant of a type that is compatible with the catalyst and suitable for the lubrication

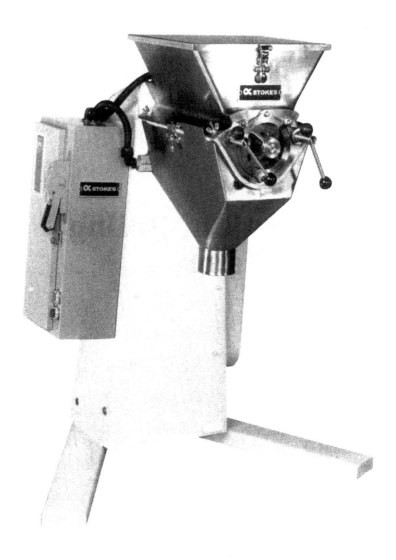

Figure 29 Stokes granulator showing exterior view with enclosed reciprocating barrel, squirrel cage. (Courtesy of Stokes-Merrill, Inc., 2 Pearl Buck Court, Bristol, PA 19007.)

Figure 30 Paul O. Abbe commercial size ball mill. (Courtesy of Paul O. Abbe, 139 Center Avenue, Little Falls, NJ 07424.)

of the dies and punches in the pilling machine. Catalyst powders have varying characteristics relating to the pilling operation. As previously indicated, some compress and cohere quite easily, whereas others are very difficult to pill and may even require the addition of certain salts during the densification operation to aid the compression and molding into pellets.

The compression operation is essentially always extremely harmful to the activity of the catalyst. The pilling lubricant itself is highly objectionable, but because of the need to have a hard catalyst particle for liquid- or vapor-phase fixed-bed operation, the pilling operation is a necessary evil. Graphite is a particularly objectionable pilling lubricant but is one of the most effective in producing hard and abrasion-resistant pills.

The crystalline structure of graphite is platelike, much like a shingle roof, and during the pilling operation these plates are rubbed along the surface of the pill. On occasion graphite may be sufficiently harmful to the catalyst activity that the catalyst must, after pilling, be oxidized to eliminate at least a portion of the graphite. This also must be done very carefully, because catalysts quite often have comparatively readily reducible oxides that provide oxygen for the combustion of the graphite, which could evolve a substantial and harmful amount of heat.

If oxygen is not readily available from air, a secondary harmful effect may be experienced. That is, the oxidation of the graphite may remove the oxygen from the catalytic material itself with a deactivating effect on the catalyst. Usually, such an effect is accompanied by a color change, and the catalyst can be readily observed to have had a harmful exposure to elevated temperatures. Unfortunately, when a catalyst has been exposed to such a violent reduction, reoxidation will not generally restore its intrinsic activity. Furthermore, frequently when the graphite is burned out at an elevated temperature, the pellet will become quite soft and may even decrepitate during the burning operation. ("Decrepitate" refers to slivers breaking away such as occurs when molten glass is dropped into water.)

The quantity of graphite ordinarily used is approximately ½–1%. Many catalysts are satisfactorily pilled with ½%, whereas other catalysts may require as much as 3% for proper pilling. To emphasize once more – graphite is harmful; the more graphite, the more harm. Other lubricants will be mentioned subsequently.

The powder that is to be mixed with the pilling lubricant is placed in a ribbon-like mixer as shown in Fig. 31B, or it can be mixed in a double cone mixer, shown in Fig. 31A. Usually the mixing is conducted for approximately 10 minutes. A warning: Inadequate mixing makes for poor pilling because the graphite or lubricant is not properly attached to the catalyst granules; also, if the catalyst is mixed for too long a time, it

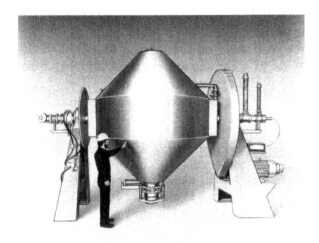

(A)

(B)

Figure 31 (A) Teledyne-Readco double cone mixer. (B) Teledyne-Readco ribbon mixer. View showing portion of ribbon spiral and charge being mixed. (Courtesy of Teledyne-Readco, 901 South Richland Avenue, P.O. Box 15552, York, PA 17405-0552.)

may self-grind and produce aerated, low-density, and comparatively nonpillable powder.

It is advisable to run tests on the mixing to establish the preferred mixing time. If after the appropriate time has been determined the rest of the operations remain uniform, then the mixing time and the pilling lubricant quantities should remain constant also.

PILLING LUBRICANTS OTHER THAN GRAPHITE

Much effort has been exerted in attempting to increase the porosity of catalysts by altering the pilling lubricant. Polyvinyl alcohol, powdered stearin, polyethylene, and other waxes or greases have been used. The quantity of these that must be used is usually two or more times the quantity that is required if graphite is used as the lubricant. On the other hand, these will oxidize and burn out of the catalyst at a much lower temperature than the graphite requires. As a consequence, the oxidization after pilling of such a catalyst can be programmed in such a way that the catalyst spends a given time (say, 30 minutes) at a given temperature, and then another 30 minutes at a higher temperature, and another 30 minutes at a still higher temperature until such time as the lubricant has been carefully burned out at comparatively low temperatures.

It has been previously stated that the combustion and removal of the pilling lubricant is a touchy operation and frequently destructive for the catalyst. However, a compensating effect is that as the organic is oxidized from the catalyst particle, voids (pores) are created where the water vapor and carbon dioxide evolve, where the organic matter was previously located. Thus, it can be seen that a benefit can be derived if the combustion is conducted carefully without thermal harm to the catalyst.

NONCOMBUSTIBLE PILLING LUBRICANTS

Some pilling lubricants, such as magnesium oxide, talc, and the like, are not combustible and thus do not introduce problems relative to combustion. However, they may be objectionable in that they are a diluent to the catalyst and may also act as a barrier to liquid or gas transport through the catalyst particle.

The objections to the use of graphite have already been enumerated except for the fact that all graphite contains a few percent of ash, and this ash can be catalytically harmful. For this reason, some of the organic lubricants are preferred although, as indicated previously, they present problems in burnout.

PARTICLE FORMATION, PELLETING, EXTRUSION, ETC.

Unfortunately, there is no agreement on the nomenclature used for catalyst forms. A catalyst particle that is formed in a pilling machine is quite often referred to as a pill, a catalytic cylinder, or a catalyst pellet. In the text so far, I have used "pill" and "pellet." Others in their nomenclature refer to pellets as those particles produced by an extrusion machine, and some even refer to the spheres made by spherudizers as pellets. In my own terminology I try to refer to extruded materials as extrudates and spheroidal material as spheres or spheroidal material.

Pilling can be done quite scientifically with the result that each particle is comparatively uniform in size, hardness, and density. In order for this to occur, however, the pilling machine must be in well-maintained condition. A pilling machine is shown in Figs. 32A and 32B. The operation and principle of the pilling machine is illustrated in Fig. 32C. A typical pilling machine as shown in Fig. 32A is manufactured by the Stokes machinery division of Pennwalt Corporation. Most catalyst manufacturers have a bank of these or a very similar machine made by Colton.

For the pilling machine to produce a satisfactory product, the dies, punches, guides, and cams must all be meticulously

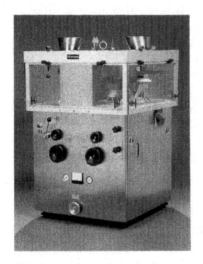

(A)

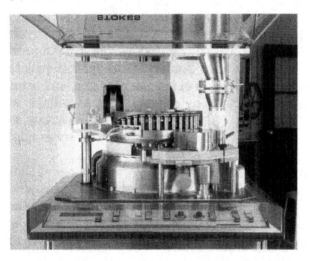

(B)

Figure 32 (A) F.J. Stokes tabletting machine. (B) Stokes-Merrill pilling machine with cover removed showing upper punches, feed mechanism, and upper compression wheel.

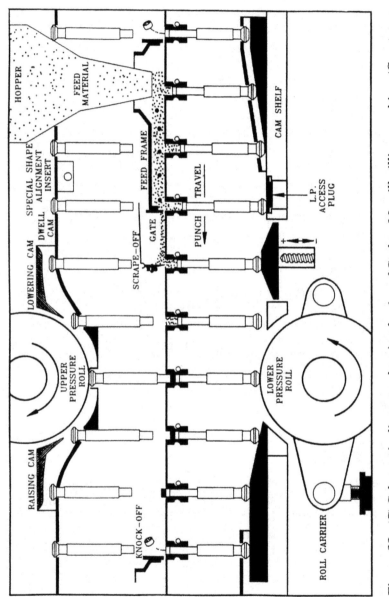

Figure 32 (C) Schematic diagram showing basics of Stokes-Merrill pilling machine. (Courtesy of Stokes-Merrill, Inc., 2 Pearl Buck Court, Bristol, PA 19007.)

maintained. Wear can cause both cracking of the pellets and disking, which is a characteristic fault when pills have been formed in a die that, because of wear, is larger where the pellet is formed than where it is ejected. Because of compression during ejection, the pellet will literally explode into chips or disks as it is ejected, and as a result the pill is of no value. It is easy to visualize that before the die reaches this point of having a very severe "disking" problem, there is an intermediate point where the pill may hold together but has strains that subsequently, in the use of the catalyst, may cause it to break into powder. Most dies, even those with carbide hardened inserts, eventually develop this barrel shape, with resultant weakening of the catalyst. Careful measurements should be made periodically on a set of dies as they are used to prevent the wear from reaching the point where the pills or pellets formed are unsatisfactory.

Usually the pellets formed are simple right cylinders of equal length and diameter. But special dies can be provided in which there is a hole through the center or the exterior is serrated in such a way as to give a corrugated appearance. Even more intricate shapes and surfaces are described in the next section under extrusion, an operation that is more amenable to these modified forms than is the catalyst pilling operation.

EXTRUSION

There are many types of extrusion equipment, some very sophisticated and expensive, but basically what they amount to is a means whereby a wet paste (or powder that is converted to a wet paste within the extruder itself) is fed to a sophisticated screw transport system and eventually emerges through dies constituting an end plate in the extruder (Fig. 28). This end plate can have holes in the shape of cogs or rings, ovals, stars, three-lobed joined rings, or hollow rings similar to macaroni. The extruder can be equipped with a slicing device that slices off the material as it is extruded.

If the extrusion is performing wet, the particles formed are very regular, hard, and uniform. However, if the extrusion is uneven and the rate of extrusion from one section of the dies is different from that of another section, the particle length can be quite variable and the hardness and sharpness of the ends can also be variable. The extruder, however, can rapidly produce great quantities of product of various shapes and as a consequence is relatively inexpensive in comparison with the pilling method of making shaped catalysts.

Although the catalysts made by extrusion generally are much less uniform and less resistant to abrasion than catalysts made in a pelleting machine, they do have better characteristics from the standpoint of porosity and freedom from lubricant, such as graphite or stearic alcohol. For a large quantity of a desiccant or a comparatively uniform support material, extrusion is a quite satisfactory production procedure.

SPHERUDIZER

Spherudizing is a special technique requiring unique equipment manufactured by Dravo Corporation, as shown in Fig. 33. In this operation the rotating disk is on an angle, and as it rotates, smaller spheres used as the seed material are placed in the bottom part of the disk and a spray of cohesive slurry is sprayed onto them. As the moisture in the slurry evaporates, the solids form a layer on the exterior of the spheres, increasing their diameter. As the spheres increase in size, they segregate into sections where the material of the desired size can be removed by specially designed draw-off techniques. There is much art in the process, first, to ensure that the build-up of the drying slurry on the pellets produces a spherical material; second, to prevent the spheres from cementing into two or three adhering spheres; and third, to avoid the formation of noncohesive layers that peel away like onion layers.

(A)

(B)

Figure 33 (A) McNally-Wellman pelletizing disc. (B) Detail of pellet pattern on 39-in. disk showing breakover point from main pan to re-roll ring.

(C)

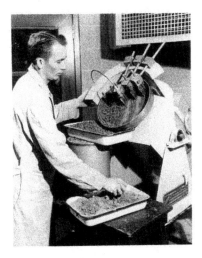

(D)

Figure 33 (C) Complete scheme of powder introduction, moisture addition, pattern of sphere, and discharge. (D) McNally-Wellman pelletizing disk of laboratory size. (Photos courtesy of McNally-Wellman, 4800 Grand Avenue, Pittsburgh, PA 15225-1599. Formerly Drano Corporation.)

The buildup of the slurry on the granules takes place in well-defined layers, and these may be strongly adhering or may be easily separated as with an onion. This onion-layering effect can be especially destructive if the spheres are to be impregnated later with a water-soluble salt. The layers may separate explosively during this impregnation; control of this problem is unique to each type of composition or fabrication procedure.

10

Spray Drying

Spray drying (Fig. 21) is another technique widely used to produce spheroidal materials. It was previously described as an intermediate process step; here we consider it as a final operation. The process described in Chapter 9 produces spheres 1/16 in. or greater in size, whereas spray drying produces microspheres that are generally smaller than 60 mesh and are useful primarily in fluidized beds. This process also is more art than science, but one who has experience in spray drying can quite well predict from viscosities, solids contents, and film-forming characteristics whether the slurry will produce suitable microspheres and, if so, what the best operating temperature, gas velocity, spray head types, and rotation speed would be. In general, it can be said that gel-forming materials – materials that shrink and coalesce on drying or that are film-forming materials – ordinarily can be spray dried to form a satisfactory product.

Many of the fluidized catalysts are processed in spray-drying equipment. Uniformity of product size, hardness, and abrasiveness are all essential characteristics for dried microspheres that are to serve as fluidized catalysts. As previously stated, the selection of specifications for the slurry for spray drying is an art that can be practices by one experienced in the process.

The types of materials that can be spray dried are typically gel-forming or coalescing slurries, but gels or film formers may also be used as matrices to occlude other more crystalline or particulate catalytic materials. For example, bismuth molybdate can be occluded in a silica gel to produce a highly satisfactory selective oxidation catalyst. In addition to the bismuth molybdate, other constituents such as phosphoric acid, iron, nickel, and alkalies can also be included. The silica gel is colloidal silica, which has a coalescing and film-forming characteristic so strong that in spite of the presence of occluded bismuth oxide, molybdenum oxide, and phosphorus oxide, a highly satisfactory catalyst—both catalytically and strength-wise—can be produced. The spray dryer and its operation are depicted in Fig. 21.

FLASH DRYING

Flash drying remotely resembles spray drying and is sometimes performed on catalytic materials in such a way as to produce a finely divided powder (Fig. 22). This powder, however, is not formed into hard spheroids; the purpose is only to remove moisture and other volatile constituents. It is obvious that flash drying can be used only when the liquid can be volatilized without leaving objectionable soluble salts under or on the surface of the devolatilized solids. Flash drying ordinarily produces a very fluffy, low-density product quite satisfactory for extrusion but insufficiently dense for use in pilling.

11

Crushing and Screening to Produce Granules

Crushing and screening constitute the procedure most frequently used on the laboratory scale to produce a granular material that can be examined in a laboratory-scale reactor (see Figs. 34–36). The way in which a powder could be moistened and densified with a mortar and pestle and subsequently densified still further by compression between two plates in a mechanic's vise was described in Chapter 2. This compressed wet cake can then be dried and the dried material crushed. The crushed dried catalyst is then screened to pass through a particular mesh size but be retained on a screen of smaller mesh below it.

Frequently one reads of granules being used that are 4–8 mesh or 12–20 mesh, and this is in general a way in which they can be derived. Some catalysts, notably the iron molybdate types used for the oxidation of methanol to formaldehyde, suffer so severely during a pilling or extrusion operation that

Figure 34   Sturtevant engineering Co., Inc., jaw crusher. (Courtesy of Sturtevant Engineering Co., Inc., 103 Clayton Street, Boston, MA 02122.)

Figure 35 Sturtevant Engineering Co., Inc. cone crusher. (Courtesy of Sturtevant Engineering Co., Inc., 103 Clayton Street, Boston, MA 02122.)

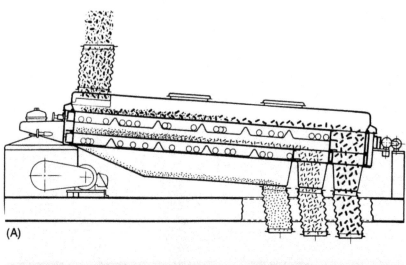

(A)

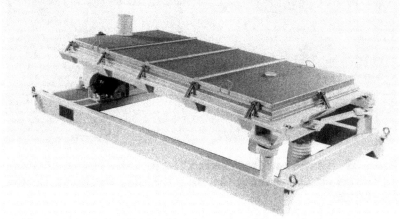

(B)

Figure 36 (A) Rotex screener. Diagrammatic view of assembled parts and flow of material into and through a screener with two screens. (B) Photograph of Rotex screener out of the normal housing and enclosure.

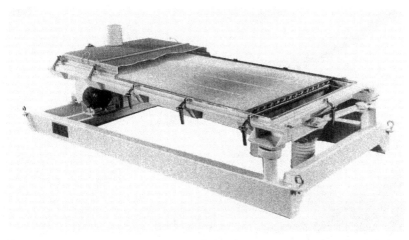

(C)

(D)

Figure 36 (C) A cutaway scheme of screen locations and product size separation and individual discharge. (D) Expanded view of major components.

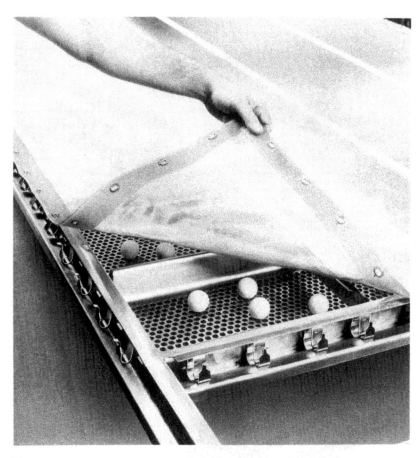

(E)

Figure 36 (E) View showing balls on the screen to eliminate blinding of screens. (Courtesy of Rotex, Inc., 1257 Knowlton Street, Cincinnati, OH 45223.)

they are processed solely to produce a mat or lump of catalyst that is then granulated and screened (Figs. 27, 29, and 36), and a useful screen size is selected for charging to even plant-scale converters. In general, it can be said that catalysts that are not further treated by reduction or dehydration or in some other way to derive the porosity will not tolerate pilling or extrusion; they must be used in the granular form or as spray-dried material.

Ordinarily, the granular material is not sufficiently hard or abrasion-resistant, and certain hardening agents are included. As examples, colloidal silica or alumina gel derived from aluminum hydroxide are frequently used. Other hardening agents are colloidal zirconia, calcium aluminate, sodium aluminate, sodium silicate, magnesium hydroxide, and aluminum hydroxynitrate. It is easy to recognize some of these as possibly being highly objectionable to the catalytic material, but often a suitable hardening agent can be derived that is satisfactory. Organic titanates can also be used; these are offered under the trade name Tyzor by the DuPont Company.

12

Coating (Not Impregnation)

Catalytic reactions are frequently conducted under conditions in which only the outer periphery of the catalyst particle functions actively. This means that the interior (frequently any portion below the outer 1000 angstroms) is catalytically of little or no value. It also is true that the site of catalytic action, which is in the lower reaches of the pores, is frequently where the objectionable by-products form and unwanted side reactions take place. As a consequence, it may be highly desirable that a relatively impervious core be coated with a thin layer (a few micrometers thick) of the catalytic material. There are also highly porous, low surface area fused supports in which the pores are quite large and can themselves be coated and because of their size will permit easy egress and ingress of the catalytic reactants.

The coating equipment generally resembles the pill-coating devices available to the pharmaceutical industry. Two of these are shown in Fig. 37. The cores to be coated are placed

(A)

(B)

Figure 37 (A) Thomas Engineering pill-coating machine. (B) Photo showing simplicity owing to the few essential parts. (C) Diagram showing flow of liquid and air to pill-coating cylinder. (A,B,C courtesy of Thomas Engineering, Inc., 575 West Central Road, Hoffman Estates, IL 60195-0198.) (D) Aubin one-quart size laboratory pill-coating machine.

110

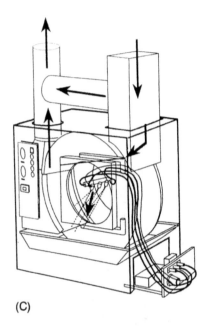

(C)

(D)

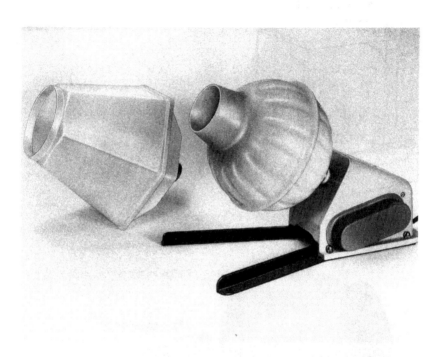

(E)

Figure 37   (E) Aubin one-gallon size laboratory pill-coating machine. (D,E courtesy of A.E. Aubin, 345 North Main St., Marlborough, CT 06447.)

in the rotating drum, and the catalytic "paint" is coated onto the core. The thickness of the layer can be controlled by the amount of "paint" slurry that is coated on the cores. It may be desirable to dry between coatings or continuously with an electric or steam jacket on the drum.

The catalytic "paint" can be compounded in much the same way as the slurry that is used in the spherudizers mentioned earlier. One of the major problems with this type of catalytic material is that the coating may peel from the core (a process called *spalling*). This has two very deleterious effects: (1) The core now becomes catalytically inert, and (2) the powder derived from the spalling will gravitate to the downstream portion of the catalyst bed, where it will very likely find a place to plug or distort the flow of gas.

It is obvious that much of the processing of coated catalyst is art, and inasmuch as it is art, it will require the typical cut-and-try methods for successful resolution of problems. The beneficial effects, however, are such as to suggest that it is worth considerable effort to find a successful solution. Not only is it possible that a more efficient and selective catalyst will be derived, but, equally important, it could be that the use of expensive materials can be very sharply reduced. A third, less obvious, benefit is that the core of the coated catalytic material acts as a heat sink and may very well tend to spread out a reaction, making it more nearly isothermal.

13

Impregnation to Orient the Coating Material to the Support

There is a growing understanding and acceptance of the fact that when they are supported on crystalline and oriented structures many catalysts assume the orientation and spacing of the support. This can be a beneficial effect and in some cases is thought to completely alter the catalytic characteristics of the composite structure. As an example, alumina in the gamma or eta form can be used as support for a catalytic material such as ruthenium, with the effect that the ruthenium behaves as a hydrogenation catalyst in an entirely different way than when it is on a different support material.

It has long been recognized that there is a very sharp difference between the performance of a catalytic material supported on gamma-alumina and that of the same catalyst on alpha-alumina, but this was usually brushed off as being attributable to the difference in the surface areas. This cannot be offered as the explanation between gamma- and eta-alumina because they frequently have very close to the same surface areas; however, they differ in crystalline structure and lattice spacing. (Eta-alumina is recognized as a spinel form.)

Usually the desired catalyst orientation is best achieved by suspending the support material in a solution of the catalytic material and then incipiently and slowly precipitating the catalytic material onto the support in such a way that it has an opportunity to orient on the support. If the precipitate is slightly soluble in the precipitating medium, then the simple expedient of digesting it for a given period of time effects the orientation. As an example, if cobalt or nickel is precipitated in an ammoniacal environment, the cobalt and nickel amines will slowly solubilize and reprecipitate, orienting on the support material.

Many other types of supports can be used, including finely divided zirconia, titania, silica, magnesia (periclase), spinels of various types such as magnesium aluminate, cobalt aluminate, and nickel aluminate, and even calcium or barium sulfates. Calcium sulfate itself is slightly soluble in high-temperature water and as a result may foil efforts to orient a catalytic material on its surface. However, by codepositing some of the support and the catalytic material, specific and unusual effects can be produced.

Coating can also be performed in a two-step operation. The support is first impregnated with a precipitant, which is then dried on the support. Subsequently the support plus precipitant is exposed to the coating technique in such a way that precipitation of the catalytic coating takes place immediately on the surface. In this scheme the coating material may be an aqueous solution of a catalytic material that precipitates as a thin film when it reaches the surface. The supernatant liquid may penetrate into the support, but as it penetrates the catalytic material remains immobilized as a thin skin on the surface. This procedure is used frequently with precious metals in order to locate them where they are most available to the reactants and are also strongly anchored to the surface.

"ANCHOR COATING" OR "WASH COATING"

On occasion it may be necessary to establish an anchor coat on denser, less penetrable support materials. The anchor coat most frequently used at present is alumina from basic aluminum nitrate as a relatively thin slurry. This anchor coating, after being placed on the support, is frequently fired or calcined at sufficiently high temperature to decompose the hydroxynitrate to form high surface area Al_2O_3. Thereafter, the support may be further impregnated by the addition of magnesium as magnesium nitrate or acetate or a slurry of the hydroxide, which will then coat over the alumina just deposited.

On firing, the magnesium aluminum oxide mixture will react to form magnesium aluminate spinel, which may give specific crystalline effects or may block the aluminum oxide lattice to prevent catalytic material from entering that lattice and becoming catalytically unavailable. Of course, other spinel-forming materials such as copper oxide, nickel oxide, titanium oxide, or cobalt oxide can be placed on the alumina to give spinels that have desirable catalytic properties and help in the proper promotion or stabilization of the catalytic material that is later placed on the support material.

Part II
Operating Procedures

14

Petroleum Processing Catalysts

The petroleum processing catalysts that have aroused the greatest interest and that constituted a major advancement when they were first introduced in the 1960s are the molecular sieves or zeolites. It was once thought that the research in zeolites had reached a mature state, but the synthesis of a continuous series of high silica/alumina materials (Silicalite and ZSM zeolites) has greatly expanded the potential of these materials. It is estimated that the annual production of zeolites is now 500,000 tons. Their major use is in detergents, with only about one-fourth of the production being attributed to use as catalyst. The preponderance of catalysts remain based on faujasite or Y-zeolite type structures. Much of the product appears to be material tailored to meet a specific need in terms of equipment or feedstock. Catalyst manufacturers are also beginning to test the usefulness of naturally occurring zeolites.

Improvements have been made in zeolites over the years, with the most recent improvement being represented by the ZSM types, which include an organic moiety that in turn introduces specific properties such as specific geometries and site conditions. Zeolites have long been a part of the mineralogy discipline; they number in the hundreds and perhaps the thousands, but only a few have been adapted for catalytic use. These vary in pore openings, alumina/silica ratio, and alkali content. When an organic moiety is present, the type of organic compound used is generally an amine or a diamine, and each type introduces unique and useful physical, chemical, and catalytic properties.

ZEOLITES

In 1756, Cronstedt coined the term *zeolite* to describe a class of naturally occurring minerals composed of silica and alumina. In 1949, Robert Milton, of Union Carbide, discovered methods to produce synthetic zeolites by a low-temperature hydrothermal process. Application of synthetic zeolites in heterogeneous catalysis has had a major impact in both the petroleum and chemical industries. The ability to modify the physical and chemical properties of the crystalline materials has led to a wide range of substances that are used in a variety of catalytic processes and in processes for selective adsorption, ion exchange, detergents, etc.

Over 150 zeolite structures have been synthesized, with compositions ranging from the original silica-alumina structures to pure silica, metallosilicates, and aluminophosphates. This review can cover preparations of only a few of the materials. The preparation of synthetic zeolites is a process of crystallization of a silica gel at temperatures above 50 °C over a varying time period. Organic modifiers are often incorporated in the gel. Due to the presence of metastable phases, excellent control of all process factors is required to obtain a single structure or in some cases, even a reproducible mixture of structures. The influence of synthesis variables has been the subject of extensive research.

The use of amines to modify the crystallization process has resulted in much speculation about the role of the organic component as a template. The observations that the presence of amines leads to new structures and that the amines are trapped in the pores of the zeolites and can be removed only by pyrolysis seemed to confirm the postulated template effect. The first synthesis of ZSM-5 described the inclusion of tetrapropylammonium hydroxide. The use of amine modifiers led to prolific research and publication, but the need to eliminate the cost and hazard of the amine additives resulted in the synthesis of ZSM-5 from an organic-free media. The presence of an organic component permits the synthesis to be operable over a wider range of gel compositions but does not appear to accelerate the rate of crystallization or improve the quality of the product.

The conversion of the hydrothermal product to a catalyst usually involves first exchanging the sodium ions with the ammonium and then calcining to the acid form. The degree of hydration may be altered, and in some cases the exterior surface of the zeolite crystals may be modified to reduce or eliminate any reactions except those governed by the internal zeolite structure. Catalysis by the zeolite crystals can be utilized directly, but in most cases the zeolite is compounded with a binder to serve as a diluent and to provide particles that conform to the process requirements of activity, pressure drop, and attrition resistance. The techniques of incorporating zeolites into a matrix are considered highly proprietary. Clays or silica continue to be the more general types of binders used in fabrication. The variation of additives, binders, and lubricants to produce an extrudate with the desired chemical and physical properties remains a critical and secretive process.

The external surface of the zeolite crystals is often modified to emphasize the particular properties of the internal crystal. Dealumination using dicarboxylic acids to reduce surface acidity can be employed. An alternative treatment with silicon compounds such as $SiCl_4$ or $Si(OEt)_4$ can lead to dealumination or to altering the Si/Al surface ratio with a modification of the surface acidity and possibly the opening of the pore mouth. The use of successive treatments of sodium aluminate and ammonium nitrate to add alumina to the structure to increase the acidity has been described.

A semicontinuous preparation of zeolites (ZSM-5) has been described. The process involves a continuous feed of pH-adjusted aluminum sulfate and sodium silicate onto an agitated autoclave with a residence time in the range of 30 minutes. The autoclave effluent is washed on a belt filter (Fig. 19) and flushed into a charge tank that allows addition of reagents and zeolite seed crystals. The charge tank material is then pumped to a batch crystallizer autoclave.

Evolving processes for the synthesis of zeolites are generally aimed at lowering the cost of manufacture and/or improving activity. The synthesis of ZSM-5 by the kneading of a dense nonaqueous system of aluminosilicate gel and sodium hydroxide in ethylenediamine and triethylamine has been disclosed. Studies on the application of ultrasonics to modify the crystallization process do not so far indicate any improvements proportionate to the expense of treatment. However, continuous improvements in the crystallization process and modification of crystal morphology are reasonably anticipated.

The literature describing the preparation of zeolites is voluminous and often rough in critical details. A large part of the open literature is from academic sources that are more concerned with characterization than synthesis. Patent literature disclosures are sometimes difficult to interpret or reproduce. A series of representative preparations has been selected to cover a wide range of zeolites. These preparations cannot be guaranteed but have been found to be reasonably reproducible. The use of organic templates is recommended for laboratory-scale preparations as they define a resilient and tolerant system. The final products are defined by X-ray crystallography and physical properties such as sorption.

The zeolites were generally produced in nature under relatively high temperature and pressure hydrothermal conditions. These, of course, can be duplicated in the laboratory but at considerable expense, so most zeolites are produced in the laboratory by crystallization under relatively mild conditions, that is, at temperatures in the neighborhood of 100–150°C and autogenous pressure. Because of the milder conditions, the time for the development of the crystallinity in the molecular sieves is usually on the order of hours or even days. The basic principle of the synthesis is shown in Fig. 4 and is described as follows:

1. An agitated solution is prepared containing sodium silicate and sodium aluminate in the proportion necessary for the final composition, which in this case let's say is 1 alumina for each 3–30 silicas. There is added to this solution sodium hydroxide to the extent necessary to achieve a preselected sodium/alumina/silica ratio in the final product. Additionally, solid finely divided silica in the form of pulverized glass sand or colloidal silica or optionally finely divided alumina hydrate is suspended in the agitated medium. The temperature is raised to near boiling (135–150°C), and the solution slurry is agitated rapidly so as to keep not only the finely divided raw materials in suspension but also the granules of molecular sieve as they are made. After the digestion period and the silica and alumina, which are introduced as the hydrate or silicon dioxide, have been converted to the zeolite, the digestion is stopped and the particles of zeolite are separated from the nutrient liquor.
2. The zeolites are unsatisfactory in the form in which they are present at this stage and are generally added to a clay such as halloisite, montmorillonite, or attapulgite or to a gel of amorphous silica-alumina. After acid washing of the clay to remove impurities, the clay may also be alkali-treated to partly convert it to a zeolite-like structure during which it becomes gel-like and cohesive, which is highly desirable

for the next step in the preparation. Approximately 10–20% of the molecular sieve is now mixed into the clay in such a way as to make a uniform paste. This is generally done in simple mixing equipment such as that shown in Figs. 27 or 28. Instead of clay, silica-alumina catalysts of the amorphous variety may be used while still in the gel stage, but clays are used most frequently as the matrix. After the clay-zeolite has been made uniform, it is spray dried as a slurry using equipment as depicted in Fig. 21A and B to form a powder.

3. The powder is in the form of 60–200 mesh spherulites and may be used in this form after ion exchanging from the alkali to acid form and finally to a rare earth form as later described.

4. The molecular sieves in this form are not satisfactory for petroleum processing, and the sodium must be removed by ionic replacement. This is done by ion exchanging the sodium, replacing it with ammonia from ammonium chloride or ammonium nitrate. Approximately 90–95% of the sodium is replaced by slurrying the spherulites in a solution of 5–10% ammonium nitrate or ammonium chloride repeatedly until sodium no longer appears in the exit stream.

5. After the sodium has been replaced with ammonium, the catalyst can be heated at 300–400°C to remove the ammonium ion, giving an acid site where the ammonium ion earlier replaced the sodium.

In this form, the catalyst is extremely acidic, so it is not applicable for cracking operation, but it becomes applicable if, instead of using it in the acid form, at least some of the acid sites are replaced with rare earth ions such as cerium, lanthanum mixtures, or others preferred by the manufacturer and user. After replacement of the acid sites with the multivalent rare earths, the catalyst is much more stable at high temperatures and is extremely effective for petroleum processing, particularly for the formation of gasoline fractions.

Molecular sieves are useful for many catalytic applications, particularly for petroleum processing, but are also finding use as detergents, desiccants, etc. Additionally, in the recent literature many references are made to the replacement of some of the acid sites with copper, manganese, nickel, iron, and other metals, which then can be used for certain other catalytic processing. An effort was made to replace the acid site with zinc and copper to produce methanol, which then, in the presence of the molecular sieve, would be converted to dimethyl ether and subsequently into gasoline fractions. To date, this has not been reported to be a satisfactory system. However, certain replacements of the acid sites by metals that catalyze other operations such as hydrogenation have enjoyed success.

PREPARATION OF ZEOLITE A

A solution of 89 g of sodium aluminate (44.1 wt % Al_2O_3, 30 wt % Na_2O, balance water) in 320 cm³ of water is prepared, and 126 g of sodium silicate solution (7.5 wt % Na_2O, 25.8 wt % SiO_2) is added with agitation. The clear solution is held at 100°C in a closed container for 12–14 hr and the crystalline product is removed by filtration at room temperature, washed to a pH of 10, and dried at 100°C.

Another method that indicates the range of compositions that produce zeolite A is to prepare a solution of 15.0 g of sodium aluminate (composition as above) in 75 cm³ of water, add 23.2 g of sodium silicate solution (composition as above) with agitation, and hold it at 100°C for 1–2 hr. The product is isolated as above.

PREPARATION OF ZEOLITE X

A solution composed of 10 g of sodium aluminate (44.1 wt % Al_2O_3, 30.0 wt % Na_2O, balance water), 5.5 g of NaOH, and

135 cm³ of water is prepared, and a 32-g portion of sodium silicate solution (20 wt % Na$_2$O, 32 wt % SiO$_2$) is added with agitation to produce a clear solution. The solution is held at 100°C for 2 days to induce crystallization. The product is removed by filtration, washed to a pH of 10–11, and dried at 100°C.

Colloidal silica may be used as a silica source. A solution of 10 g of sodium aluminate, 29.7 g of sodium hydroxide, and 72 cm³ of water is prepared, added with vigorous agitation to a 228-g portion of colloidal silica (29.5 wt % SiO$_2$, Ludox) to give a clear suspension, and then aged at 50°C for 10–14 days. The product is recovered as above. This illustrates the resilience of crystallization to produce zeolite X over a wide range of compositions.

PREPARATION OF ZEOLITE Y

A solution composed of 5 g of sodium aluminate (by weight 44% Al$_2$O$_3$, 30.0% Na$_2$O, 25.9% H$_2$O), 22 g of sodium hydroxide, and 89.5 mL of water is prepared, and a 124-g portion of colloidal silica containing 29.5 wt % SiO$_2$ (DuPont Ludox) is added to give a composition with a molar ratio of 13.9 Na$_2$O: 1Al$_2$O$_3$:28.2 SiO$_2$:471 H$_2$O. The mixture is stirred for 2–4 hr until homogeneous and then held at 100°C for 21 hr to induce crystallization. The product is removed by filtration, washed to a pH of 10–11, and dried at 100°C.

Alternatively, the homogeneous mixture may be crystallized in an autoclave at 120°C, which reduces the time to 3 hr.

PREPARATION OF SILICALITE

A mixture is prepared by adding a solution of 1.4 g of NaOH in 10 mL of water to 44 g of aqueous colloidal silica (30 wt % SiO$_2$, Ludox) with stirring. A solution of 2.4 g of tetrapropylammonium bromide in 15 g of water is then added with continuous agitation. The synthesis mixture is placed in a Teflon-lined pressure vessel and held at 200°C for 3 days. The solid reaction product is recovered by filtration, water washed, and dried at 105°C. It is then calcined in air at 600°C to remove the entrapped organics.

Alternative preparations using tetrapropylammonium hydroxide may be substituted, which eliminates the use of sodium hydroxide and bromide. A solution of 9.0 g of tetrapropylammonium hydroxide in 25 g of water is combined with 44 g of aqueous colloidal silica (30 wt % SiO_2). The crystallization and product recovery are carried out as above.

Very fine silica, such as Cab-O-Sil fumed silica, may be substituted for aqueous colloidal silica. A solution of 12 g of tetrapropylammonium hydroxide in 30 g of water is combined with a suspension of 20.8 g of Cab-O-Sil silica and processed as above to produce silicalite.

PREPARATION OF ZSM-5 ZEOLITE

A 22.9-g portion of fine silica is dispersed (partially dissolved) in 100 mL of 2.18 M tetrapropylammonium hydroxide by heating to 100°C. A solution of 3.19 g of $NaAl_2O_3$ (analysis 42 wt % Al_2O_3, 30.9 wt % Na_2O, 27.1 wt % water) in 53.8 mL of H_2O is prepared. The resulting mixture is aged in a Pyrex-lined autoclave at 150°C for 6 days. The crystalline product is washed and dried at 105°C and then calcined in air at 620°C for 16 hr.

A preparation of ZSM-5 zeolite that avoids the use of seeds or templates has been disclosed in European patent 0098641. A mixture of 100 g of Na_2SiO_3, 6.3 g of $Al_2(SO_4)_3$, and 250 g of sodium silicate by weight 260.3% SiO_2, 8.92% Na_2O, and 65% water) and aluminum sulfate [$Al_2(SO_4)_3 \cdot 16\ H_2O$ assaying 17 wt % Al_2O_3] is held at 175°C in an autoclave with an agitator for 11 days to produce pure ZSM-5.

PREPARATION OF ZSM-11 ZEOLITE

A 100-g portion of colloidal silica (DuPont Ludox, 29.5 wt %
silica) is added to a solution of 1.2 g of sodium aluminate (41.8%
Al_2O_3, 31.6% Na_2O, balance water), 3.5 g of sodium hydroxide,
35.2 g of tetrabutylammonium iodide, and 170 g of water with
agitation for 10 min and the cloudy product is maintained at
100°C for 20–25 days. The crystalline product is removed by
filtration, water washed to a pH of 7–8, and dried at 110°C.
The dry product is calcined in air at 625°C for 24 hr. The final
Si/Al ratio is about 80.

PREPARATION OF ZSM-20 ZEOLITE

A solution composed of 14.1 g of sodium aluminate (analysis
by weight 43.5% Al_2O_3, 30.2% Na_2O, 24.9% H_2O), 764 mL of
1.51 M tetraethylammonium hydroxide, and 6.0 g of a 50%
aqueous sodium hydroxide solution is vigorously mixed, a
282-g portion of tetramethylorthosilicate is added over a period of 15 min, and the mixture is stirred for an additional 30–
45 min to complete the hydrolysis of the ester. The slurry is
aged for 3 days at ambient temperature and then held for 3–4
weeks at 100°C to complete the crystallization. The resulting
crystalline product is filtered and water washed to a pH of 7–8.

PREPARATION OF ALUMINOPHOSPHATE
($AlPO_4$) ZEOLITE

To a solution of 46.1 g of 85% orthophosphoric acid in 100 mL
of water is added 27.5 g of hydrated alumina (a pseudo-boehmite composition of 74.2 wt % alumina and 25.8 wt % water,
commercially available under the Catapal trade name). The
mixture is stirred to produce a homogeneous suspension, and
176.8 g of aqueous solution of tetrapropylammonium hydroxide containing 23 wt % base is added, followed by stirring for

1-2 hr until homogeneous. The resulting suspension is transferred to a Teflon-lined autoclave and treated at 150°C under autogenous pressure for 48 hr. The solid reaction product is recovered by filtration, water washed, and air dried at ambient temperature. Calcination in air at 1000°C for 4 hr removes the organics to produce $AlPO_4$-5.

In a similar fashion, but using 20.3 g of di-n-propylamine and carrying out the hydrothermal treatment at 200°C, the zeolite $AlPO_4$-11 is produced. By variation of the organic additive, a series of $AlPO_4$ zeolites may be prepared starting with the same A/P molar ratio in the parent solution.

PREPARATION OF SILICOALUMINOPHOSPHATE ZEOLITE

SAPO-5

First 4.58 g of pseudo-boehmite (74.2 wt % Al_2O_3) is dissolved in a solution of 7.7 g of 85% orthophosphoric acid in 33.3 g of water. To this solution is added 2.16 g of fumed silica (92.8 wt % SiO_2) and the mixture is stirred until homogeneous. Then 16.3 g of a solution of 40 wt % tetraethylammonium hydroxide is added, again followed by stirring until homogeneous. The reactor mixture is placed in a Teflon-lined autoclave and held at 150°C for 4 days. The reaction product is recovered by filtration and washed and dried in air, followed by calcination in air at 550°C for 24 hr. An alternative preparation may use colloidal silica as the silica or aluminum isopropoxide as the alumina source.

SAPO-11

To a solution of 90.79 g of aluminum isopropoxide and 51.3 g of 85 wt % orthophosphoric acid in 160 g of water is added 1.4 g of fumed silica. With good agitation, 22.2 g of di-n-propylamine is added. The reaction mixture is then heated at 150°C for 3 days in a Teflon-lined autoclave. The product is recovered as above.

TITANIUM SILICALITE TS-1

To a solution of 45 g of ethyl orthosilicate and 15 g of ethyl orthotitanate in a CO_2-free atmosphere is added over 4 hr 800 g of a 25 wt % solution of tetrapropylammonium hydroxide, which must be free of inorganic base. After 1 hr of stirring the temperature is raised to 90°C over a 5-hr period to distill off the ethanol produced by hydrolysis. Distilled water is added to give a total volume of 1.5 liters, and the resulting sol is transferred to a titanium autoclave and held at 175°C at autogenous pressure with stirring for 10 days. The autoclave is cooled to ambient temperature, and the product is recovered by filtration. The product is washed with an equal volume of hot distilled water at least four times, dried at 110°C, and then calcined at 550°C.

Alternatively, titania gel (freshly prepared by hydrolysis of ethyl orthotitanate and converted to the peroxytitanate) can be employed. A 150-g portion of ethyl orthotitanate is slowly added to 2.5 liters of distilled water over 2–4 hr with agitation. The sol is cooled to 5°C, and 1.8 liters of 30% hydrogen peroxide is precooled to 5°C and added with mild agitation over a 2-hr period with additional cooling to maintain 5°C. To the resulting clear orange solution is added 2.4 liters of 25% aqueous tetrapropylammonium hydroxide, also precooled to 5°C. After 1–2 hr, 500 g of Ludox AS containing 40% colloidal SiO_2 is slowly added with mild agitation. The mixture is allowed to stand for 12–18 hr, heated with stirring to 70–80°C, and held for 5–7 hr. It is then transferred to a titanium autoclave and held at 175°C under autogenous pressure for 10 days with stirring. The product is recovered as above. This procedure is particularly amenable to variation in the Ti/Si ratio.

Increased acidity may be incorporated by the addition of aluminum as sodium aluminate to the Ludox. Similarly, boron may be incorporated by adding potassium borate to the Ludox portion.

The applications of TS-1 zeolite for highly specific oxidations with hydrogen peroxide are described in U.S. patents 4,480,135 and 4,833,260.

An earlier disclosure (U.S. patent 3,329,481) appears to describe the preparation of a closely related material. The preparations describe the use of zirconyl chloride or titanyl chloride, which is converted to the soluble sodium peroxotitanate using caustic and hydrogen peroxide. After decomposition of the excess peroxide, silica is added as sodium silicate, the solution is heated briefly for 30 min at 100°C, and on cooling neutralization with acetic acid precipitates sodium silicotitanate.

The preparation of a series of metallosilicalite aerogels has been described by Neuman (*J. Chem. Soc. Chem. Commun.*, *1993*,1865–1867).

PREPARATION OF BOROSILICALITE AMS-1b ZEOLITE

Tetra-*n*-propylammonium bromide (9.4 g) is added to a solution of 0.25 g of boric acid and 1.6 g of NaOH in 60 mL of distilled water, and the mixture is stirred until the tetra-*n*-propylammonium bromide is dissolved. A 12.7-g portion of Ludox AS (30% silica) is now added with vigorous stirring to produce a gelatinous mixture. The mixture is treated at 165°C in a Teflon vessel under autogenous pressure for 4 days. The crystalline product is recovered by filtration, washed with distilled water, and dried at 160°C. The organics are removed by calcining in air at 600°C for 4 hr. The sodium may be exchanged with ammonium nitrate or ammonium acetate, and after calcination of the exchanged material at 500°C for 4 hr, other metals can be exchanged into the zeolite.

PREPARATION OF CHROMOSILICATE AMS-1 Cr ZEOLITE

To a solution of 0.7 g of $Cr_2(SO_4)_3 \cdot 15\ H_2O$ in 60 mL of water is added a solution of 2.6 g of NaOH in 10 mL of water with vig-

orous stirring to precipitate the flocculent chromium hydroxide. To this slurry is added 10.2 g of tetra-n-propylammonium
bromide, and the mixture is stirred until the bromide is dissolved. A 13.5-g portion of Ludox AS (30% silica) is added with
vigorous stirring over 1–2 hr. The resulting green suspension
is placed in a Teflon-lined autoclave and held at 165 °C under
autogenous pressure for 7 days. The resulting crystalline solid
is recovered by filtration, washed with distilled water, and air
dried at 165 °C. Recovery gives about 2 g of material. The organics are removed by calcination in air at 500–600 °C.

A number of substitutional variations are possible by ion
exchange. Compositional variations of the original material
are reported to be produced by cycling, oxidizing, and reducing calcinations.

PREPARATION OF A NU-10 TYPE OF ZEOLITE

A 38 g portion of carbonate-free tetraethylenepentamine is
dispersed in 142 g of colloidal silica (Syton X-30, Monsanto),
and then a solution of 1.6 g of sodium aluminate (Na$_2$O, Al$_2$O$_3$
·2.4 H$_2$O) in 10 mL of water is added to the silica suspension.
A solution of 23 g of NaCl in 388 mL of water is finally added.
This should give a composition with a molar ratio of 2.34 Na$_2$O:
27 TEPA:1 Al-O$_3$:96.6 SiO$_2$ with 3878 H$_2$O and 54.8 NaCl.
The gel is stirred for ½ hr at 20–25 °C and then held for 3 days
at 180 °C in a stainless steel autoclave at autogenous pressure
with stirring at 400–600 rpm. The autoclave is cooled to 60 °C,
and the product is filtered and washed.

The washed material is calcined in moist air at 450 °C for
48 hr steeped in 1 N HCl, washed, and dried at 120 °C. If amine
(C analysis) is still present, the material is once again calcined
in moist air at 450 °C for 4–8 hr.

The resulting product should have a SiO$_2$/Al$_2$O$_3$ ratio in
the range of 85 and acid and X-ray pattern as specified in the
reference. The Al may be at least partially replaced by Fe, Cr,
B, V, Mo, As, Mn, or Ga.

The aqueous colloidal silica may be replaced with an aerosol silica, and the tetraethylenepentamine may be replaced by ethylenediamine, but such alteration requires a change in formulation and/or crystallization profile.

REFORMING CATALYSTS

Reforming catalysts employed in the petroleum industry were probably the first to be recognized as dual functioning. By dual functioning is meant that the catalyst will perform both cracking operations, that is, it will convert large molecules of petroleum into smaller molecules but simultaneously, because of acid site characteristics, it will isomerize these new fragments into aromatic or branched-chain materials having high octane ratings. There is generally on the catalyst, in addition to the cracking function, a hydrogenation function such as platinum and rhenium or manganese that gives sulfur tolerance and minimizes carbon deposition. There is a great deal of theory that does not lie within the scope of this book but would be interesting background for anyone involved in the manufacture and application of this catalyst. (See the references.)

REFORMING CATALYST MANUFACTURER

The first step in the preparation of the reforming catalyst is the manufacture of the support. It is essential that highly acidic support be prepared, and this can be achieved by precipitating alumina from aluminum chloride in such a way that chlorine is retained in the aluminum oxide. This retained halide produces an acid function that is relatively stable. This function can be introduced into the aluminum oxide derived from other sources, however, by intentionally exposing the aluminum oxide to chloride or halide vapor. We will describe the process only as it relates to the preparation of the alumina from aluminum chloride, whereas the maintaining of the chloride during use in the catalytic structure by the continual addition of an inorganic or organic compound will be described and this method could be used similarly for chlorinating and introducing acid sites into alumina derived from other sources.

Step 1. An aqueous solution is prepared by dissolving sufficient aluminum chloride in distilled water to produce a 10% solution. Ammonium hydroxide is added to this solution in sufficient quantity to increase the pH to approximately 7 (neutrality). The aluminum hydroxide formed is then separated from the liquid, generally by spray or flash drying (Figs. 22 and 23); filtration can be employed but is extremely slow for this gelatinous product.

Step 2. The slurry is spray dried to produce alumina with a substantial amount of retained ammonium chloride or halide in the form of an aluminum oxyhalide. As a consequence, this is a highly acidic alumina.

Step 3. Dried product from step 1 is slurried in distilled water to remove the halide in excess of that necessary to produce the required amount of acid sites in the alumina support. This washing can be carried out in tanks such as the one depicted in Fig. 9.

Step 4. The alumina is now filtered by any of the means shown in Figs. 12–14, 19, or 21, depending upon the size of the production, whether semiworks, laboratory, or plant scale.

Step 5. The filter cake can at this point be made into a slurry and spray dried or made into a paste to be extruded. If it is extruded, this can be done in equipment such as that shown in Fig. 28. If it is to be spray dried, this can be done in equipment such as that shown in Fig. 20. If the catalyst is spray dried, then it can also be formed into particulate materials using a pelletizing machine such as the one shown in Fig. 32 or spherudizers such as the one shown in Fig. 33.

Step 6. The alumina is impregnated with platinum and rhenium by first placing the alumina in a coating device such

as those shown in Fig. 37A,B,C, or D. Sufficient platinum
chloride solution is then sprayed onto the support to provide approximately 0.4–0.5% platinum. After the coating
with the platinum chloride, the catalyst is either dried in
the pill-coating machine, which can be equipped with a
steam jacket, or separately dried in trays in equipment
such as that shown in Figs. 22 and 26.

Step 7. After the platinum has been applied and dried unto
the surface, a solution is prepared having sufficient rhenium to give a rhenium content of between 0.1 and 0.5%.
The rhenium salt can be either ammonium perrhenate or
perrhenic acid.

Step 8. The perrhenic acid solution is sprayed onto the plat-
inum-impregnated alumina support in such a way that it
is intimately associated with the platinum and the alumina.

Step 9. The catalyst is dried and is then ready for evaluation
or use. It is necessary to reduce the rhenium from the hep-
toxide to a lower state of oxidation during the early part
of the use of the catalyst or as a separate reduction operation before the catalyst is put into operation.

Step 10. During use there is a tendency for the catalyst to
lose the halide, and as a part of the halide replacement
scheme, a freon fluorocarbon such as a fluorochloro hydrocarbon or HC1 or HF is either continuously or periodically added to the feed to sustain the acidity of the catalyst, which otherwise would be lost by volatilization or
sublimation. However, the halide, which is relatively loosely
retained by the alumina, can readily be controlled continuously at a desired level by this replacement technique.

Step 11. It may be desirable to introduce into the support
pores that are much larger than those produced normally
during the fabrication of the support. This can be accomplished by introducing into the alumina support materials
that are either volatilized, oxidized, or melted out of the
structure. These can be fibrous materials such as cotton
linters or synthetic fibers: granular material such as sawdust or cork powder; easily melted material such as par-

affin wax, steric alcohol, polypropylene, polyethylene, or polyvinyl alcohol; or cellulosic fibers such as paper pulp or the commercial material Avicel. This increase in porosity is generally useful in catalysts used to process crude that is high in metals or other ash.

HYDROTREATMENT CATALYSTS

Hydrotreatment catalysts are presently going through a very major development program. The primary reason for this development program is that the catalysts presently used for hydroprocessing are quite satisfactory when employed for the lighter crude petroleum, but the United States and much of Europe may find the lighter petroleum crude to be unavailable either because the supplies are exhausted or because of geographic or political problems. As a consequence, the need to develop processes that are satisfactory for heavier crudes, coal-derived liquids, and shale and tar sands oil is evident. The catalysts that have been used in the past for hydrotreating lighter crudes are unsatisfactory for the heavier, more refractory high-ash crudes.

Hydrotreatment involves primarily the removal of nitrogen and sulfur from the crude oil, but oxygenated products are also simultaneously deoxygenated. It is necessary to remove these impurities because of their adverse effect on the catalysts used in the subsequent operations and also because it would be harmful to the environment for this type of crude to be used in combustion or in mobile equipment. The hydrotreatment catalysts generally fall in the category of a support such as alumina impregnated with two types of catalytic elements, one type being nickel or cobalt and the second being molybdenum oxide or tungsten oxide. The catalyst generally is sulfided in situ prior to its use. Sulfiding converts the nickel or cobalt to the sulfide and the molybdenum or tungsten oxides to the disulfides. The procedure for the preparation of the catalyst is as follows:

Step 1. A 10% solution of aluminum chloride is made using distilled water as a solvent. This solution is rapidly agitated both during the solution and in the subsequent precipitation step. This solution must be prepared in a tank constructed of a chloride-resistant material such as wood, glass fiber, reinforced epoxy resin, Haveg, ceramic, or polypropylent. A typical glass-lined tank is shown in Fig. 9.

Step 2. With the aluminum chloride solution at 50°C, ammonium hydroxide is added rapidly while the solution and slurry are being rapidly agitated, to the point where the pH reaches essential neutrality.

Step 3. The slurry is then preferably spray dried using the equipment shown in Figs. 20 and 21, or it can be filtered and washed using rotary or Neutsche filters such as are shown in Figs. 12–15.

Step 4. The spray-dried product, if the procedure above was used, is mixed with water together with nitric or hydrochloric acid to produce small amounts of aluminum oxyinitrate or oxychloride to eventually serve as a binder for the alumina. The paste that is thus produced can be extruded and processed as previously described in the extruder shown in Fig. 28 or dried in a dryer (Fig. 21). If instead of forming the dry powder a filter cake is derived, then this can be extruded directly in the equipment previously described.

Step 5. After the extrusion, the support is dried and eventually calcined at 300°C for 2 hr or longer.

Step 6. A solution is prepared containing sufficient cobalt nitrate to be equal to 3.5% elemental cobalt when deposited on the alumina extrudates of step 5.

Step 7. The alumina is placed in a pill- or extrudate-coating device such as that shown in Fig. 37 and is then impregnated with the cobalt nitrate solution prepared in step 6.

Step 8. The coated alumina is dried either by heating the pill-coating bowl or by a separate drying operation.

Step 9. The acidic alumina now coated with cobalt nitrate is impregnated by spraying it with sufficient solution of ammonium molybdate (ammonium heptamolybdate) to equal to 15% molybdic oxide on the support already impregnated with cobalt.

Step 10. The support, now impregnated with both cobalt nitrate and ammonium molybdate, is heated to 300°C to decompose the nitrate and ammonium molybdate salts and produce a cobalt oxide-molybdenum oxide-aluminum oxide mixture, which at this temperature of calcining will produce a small amount of solid-state reaction products between these three oxides.

The calcining catalyst is now ready for installation in the unit to be used for the hydroprocessing. However, before the catalyst is put in operation, it should preferably be sulfided using a gas comprising 90% hydrogen and 10% hydrogen sulfide at 300–400°C. This converts the cobalt to cobalt sulfide and the molybdenum to molybdenum disulfide. The catalyst, after installation in the unit, is ready for use in hydroprocessing, which can be effected at temperatures of 300–450°C and pressures of 500–4000 psi. Sulfiding can also be carried out in the unit and usually is.

Instead of the ammonium heptamolybdate one can use ammonium tungstate, and instead of cobalt nitrate one can use nickel nitrate. Other salts can be used, but it is typically preferred to use nitrate in the presence of the ammonium molybdate.

The alumina preparation procedure called for above produces a gamma-alumina. Other types of alumina such as eta-alumina derived from aluminum alcoholate may also be used. There are literature references to the use of eta-alumina to obtain a superior catalyst.

The support can sometimes be beneficially affected by the incorporation of a material that will oxidize, vaporize, or melt out of the "set" alumina shapes. Cotton linters, powdered activated carbon, carbon black, sawdust, powdered paraffin, polypropylene, polyethylene, and polyvinyl acetate can all be added to the support in the intermediate gel stage, and then subsequently these fugitive materials, upon removal by the previously mentioned procedures, will leave various types of macropores in the catalyst that can act as very large corridors for the reactants and products and also as reservoirs for the ash and tarry materials that will accumulate during the processing of extremely heavy crudes, coal-derived liquids, tar sands, and crude shale oil. Such types of catalysts are referred to as *demetallizers*.

The quantity of cobalt or nickel can be varied from the 3.5% stipulated above, and the molybdenum can also be varied from the 15% stipulated, but it is usually good practice to maintain approximately a 5-to-1 relationship between the molybdenum and the nickel or cobalt.

USED OR SPENT HYDROPROCESSING CATALYSTS— REGENERATION, RECYCLING, AND METALS RECOVERY

Regeneration

For regeneration of the used (not spent) hydroprocessing catalyst, the only condition in which catalyst can be regenerated is when it has been deactivated by carbon and is free of contaminants such as nickel and vanadium derived from certain crudes. When carbon is the deactivator, the deactivated catalyst is removed from the reactor, and the carbon is removed by carefully controlled oxidation of the carbonaceous deposit. The catalyst can be regenerated by this procedure at most twice, after which the physical changes in the catalyst render it "spent." At this point metals recovery is practiced as described below.

Metals Values Recovery

If the catalyst has been fouled by metals from the crude petroleum such as nickel, vanadium, and iron oxides, it must be

reoxidized to remove the carbonaceous deposit. Then the organic-free catalyst is treated as described below to recover
the cobalt, nickel, molybdenum, and tungsten that may be
present as catalytic components. The vanadium and nickel
contents may be present in sufficient values to justify their
recovery as well.

Recovery of Metallic Values

Spent catalyst is returned to one of the processors who handle
45,000 tons of spent catalyst that become available each year.
After the removal of the carbonaceous contaminant, the spent
catalyst is reacted with alkali to convert the WO_3 and MoO_3
(and vanadium, if present) to water-soluble alkali salts. These
salts are separated by ion exchange, and the remaining solid
is heated in an arc furnace to segregate the alumina. The nickel
and cobalt are converted by a wet treatment to the ammine
for recovery, and the alumina is disposed of as an inert slag.

   Safety Note: All spent catalysts must be handled by properly trained persons protected by suitable clothing and breathing equipment.

ALKYLATION CATALYSTS

Alkylation refers to the process typified by that in which iso-
butane is added to butene to produce isooctane or triptane.
Triptane is the isooctane that is the standard of 100 for octane
rating. Alkylation can also be used for the addition of ethylene
to benzene.

   The catalyst that is used is a highly acidic material. It can
be a homogeneous catalyst such as hydrogen fluoride (HF) or
aluminum chloride, or a solid catalyst such as diatomaceous
earth impregnated with P_2O_5, HF, or HCl. Silica gel can be
similarly used but is usually more expensive and difficult to
make. The acid is slowly sublimed, so a continuous flow of the
HF, HCl, or halocarbon is necessary to maintain the acidity

of the catalyst. P_2O_5 as phosphoric acid and SO_3 as H_2SO_4 can be used in place of HF and HC1. In the case of sulfuric acid, one must always keep in mind that the sulfur may be converted in the presence of carbon to carbon disulfide or COS.

The procedure for preparing the alkylation catalyst is quite simple:

Step 1. Pelleted or granular diatomaceous earth is charged to a coating device such as those shown in Fig. 37A,B,C, and D.

Step 2. Sufficient 10% aqueous solution of phosphoric acid is prepared to equal 10% P_2O_5 on the silicious support charged to the coating device.

Step 3. The phosphoric acid solution is sprayed onto the support, and the coating device is allowed to rotate for 10 min while the liquid is uniformly absorbed by the silica. There should be no excess liquid, but if there is, it is satisfactory if a jacketed pill-coating device is used and heated to a sufficiently high temperature that all excess liquid is either evaporated or absorbed onto the granules.

Step 4. The P_2O_5-impregnated support is now dried under carefully controlled conditions to deposit the P_2O_5 onto the support. After drying, the catalyst is ready for use in the alkylation system.

Instead of the phosphoric acid, a similar quantity of sulfuric acid, aluminum chloride, or hydrochloric acid can be used.

Instead of the silicious support, one can use titania or zirconia. The latter are more difficult to prepare or to obtain commercially, but titania as a support has recently appeared in the marketplace, offered by a Japanese concern.

15

Catalysts for Synthesis Gas Processing

In the preparation of synthesis gas there are at least four different catalytic operations. Each of these catalyst preparations will be discussed separately and in the chronological order in which the gas sees the catalyst.

In the processing of synthesis gas, pressure drop may be a major problem, and if so it becomes necessary to minimize it by suitable alterations to the catalyst structure. This minimization of pressure drop may be required to avoid increased capital or operations costs occasioned by equipment and energy requirements to repressure the gas. Because of these considerations it may be necessary to place the catalysts on honeycomb structures. Instructions on how to apply the catalysts to honeycombs are given below.

SULFUR REMOVAL CATALYSTS

There are two basic types of sulfur removal catalysts. The first is purely a chemical reagent and comprises zinc oxide, which is converted to sulfide during the course of the operations. The second is copper or iron oxide on activated carbon. The latter two are generally regenerated in place with steam. The preparation of these catalysts is quite simple and is described in the following paragraphs.

Zinc Oxide

Step 1. The simplest method for preparation of this catalyst is to purchase zinc oxide in powder form, mix it with a pilling lubricant, and pill it to form catalyst particles of the size needed in the operation. Generally, these are 3/16 × 3/16 in. in size. The purchased zinc oxide can also be made into a paste and extruded as previously described for other operations. The type of pilling machine used in these operations is shown in Fig. 32, and the extruder is shown in Fig. 28. The paste can be formed by mixing the powder with water or with water plus a small amount of acid (10% of stoichiometric for conversion completely to nitrate); either acetic or nitric acid would be satisfactory. The purpose of the acid is to assist in the strengthening of the pills or the extrudate during the eventual calcining operation.

Step 2. After the pellets have been formed, they generally are heat treated at approximately 350–400°C to increase the porosity of the catalytic particle. The porosity of the pellet or particle can also be increased by using an organic pilling lubricant such as stearin (trade name Sterotex) or fibers such as cotton linters or materials such as polypropylene, polyethylene, or polyvinyl acetate that will melt away, leaving a void.

Step 3. If the zinc oxide is to be placed on honeycomb, a slurry of the zinc oxide and film former (colloidal silica or ceria or zinc nitrate, for example) in which the film former is

present at a content of 2-10%. The decomposing film former acts as the adhesive or binder. The slurry is made in
sufficient volume that it can be sprayed onto the honeycomb or the honeycomb can be immersed in the slurry
and allowed to drain, after which it is dried. Spraying or
immersing are repeated until the desired amount of zinc
oxide has been applied.

Copper or Iron on Activated Carbon

Activated carbon, in either the granular or extruded form, is
placed in pill-coating device (Fig. 37). The activated carbon is
now impregnated with either copper acetate or iron acetate
so as to produce a content of approximately 8-15% elemental
metal, which converts to the oxide after calcining. Calcining
is effected in an inert atmosphere at a temperature of approximately 250°C to decompose the acetates. The catalyst must
also be cooled in an inert atmosphere; otherwise it will oxidize
if exposed to the air and thus be destroyed as well as creating
a hazardous situation.

If the copper oxide or iron oxide is to be supported both
on carbon and on honeycomb, the following is a satisfactory
procedure. Purchased or precipitated finely divided iron or
copper oxide and powdered carbon are placed in a ball mill or
similar apparatus. A slurry is prepared in the mill by adding
the desired proportion of iron or copper oxide and powdered
activated carbon of the type used in water purification to the
mill with sufficient water to produce a paintlike suspension
after milling. Also added to the mill initially is colloidal silica,
colloidal ceria, or other film former, and the mixture is milled
for 2 or more hours. Either the milled paintlike slurry is sprayed
onto the honeycomb structure or the honeycomb is immersed
in the slurry, drained, and dried. After drying, the spraying
or immersing is repeated until the coating weighs about 50%
of the weight of the original honeycomb. Because of the carbon content of the film, drying and calcining are performed
in an inert atmosphere.

STEAM HYDROCARBON REFORMING CATALYSTS

In steam hydrocarbon reforming process, steam plus methane or some other hydrocarbon is passed over a catalyst at elevated temperatures and elevated pressures to convert the steam and the hydrocarbon to carbon monoxide, carbon dioxide, and hydrogen. In ammonia synthesis, gas from the primary reformer (the effluent) is mixed with air and passed at still higher temperatures over a catalyst for secondary reforming of the last of the hydrocarbon and also to introduce the necessary nitrogen to synthesize ammonia. Two types of catalysts are involved in this operation.

Primary Reformer Catalyst

The primary reformer catalyst is usually alumina and nickel oxide. When operating with naphtha, a small amount of potassium, rhenium, or manganese or all three may also be added to minimize carbon deposition. The primary reforming catalyst can be made by two procedures: by coprecipitation of the nickel and the alumina or by impregnating alumina pellets or rings with nickel nitrate, which is subsequently converted to nickel oxide. The procedure for the manufacture of the coprecipitated catalyst is as follows:

Step 1. Dissolve sufficient nickel nitrate in distilled water at 70°C to form a 1 M solution. Also an amount of alumina hydrate in the form of $\frac{1}{2}$-μm or smaller particles for the Al_2O_3 to be equal to between one and three times the weight of the nickel in the solution. Note that the calculation of the relationship between weights of the nickel and alumina to be used should be on the basis of elemental nickel and aluminum oxide, not alumina hydrate.

Step 2. With the slurry being agitated at 75°C, add sufficient sodium bicarbonate to increase the pH to approximately 7.2. This completely precipitates the nickel in and on the alumina hydrate as basic nickel carbonate. The precipitation vessel is of the type shown in Fig. 9. Allow the slurry to agitate for 30 min after it reaches the desired pH. Filtration can be started immediately after this 30-min period in either of the types of filters shown in Figs. 12–15. The slurry should filter easily, and it should be washed completely to remove the sodium by proper selection of the filter means and adjustment of the washing. If adequate removal cannot be obtained by this means, the filter cake must be reslurried, refiltered, and rewashed on the filter. It is axiomatic that sodium must be avoided in catalysts that operate at temperatures required for the steam hydrocarbon reforming reaction (ca. 800°C).

Step 3. After filtering and washing, the catalyst is dried in equipment such as that shown in Fig. 18.

Step 4. After drying, the catalyst is then calcined at 400°C in equipment such as that shown in Fig. 17, 18, or 21.

Step 5. After calcining, the catalyst is densified using kneaders of the type shown in Fig. 20. The kneaded catalyst is dried in equipment such as that used in step 3.

Step 6. The catalyst is granulated (Fig. 29) to form particles that will pass through a 10 mesh screen.

Step 7. The granulated powder is mixed with a pilling lubricant, which can be either 1% graphite by weight of powder or 3 or 4% organic lubricant such as stearin (Sterotex) based on the weight of the powder. The mixing with the pilling lubricant is effected in a ribbon mixer (Fig. 31).

Step 8. The catalyst is now pelleted in a pelleting machine as shown in Fig. 32 or is extruded in equipment as shown in Fig. 28 to form cylinders or rings. If the powder is to be extruded, it is, of course, converted to paste by a kneading operation as in step 5, without the subsequent drying and granulation steps.

Step 9. The catalyst is finally calcined at 1000–1100°C in a muffle furnace or a continuous furnace as shown in Figs. 22 or 23. After calcining, the catalyst is ready for use.

In some cases when the catalyst is to be extruded or even in some cases when it is pelleted, a cement such as calcium

aluminate is added during the kneading operation to an extent
of between 5 and 15% to increase its hardness after the calcining operation. This makes possible the use of lower concentrations of nickel, which otherwise is the cementing or bind-
ing agent. However, calcium aluminate has a deleterious effect
on the catalyst activity. It should also be noted that in addition to the alumina other refractory oxides such as thoria or
magnesia can be used, but silica should not be used. Under
the operating conditions of the steam hydrocarbon reforming
reaction, silica tends to migrate from the catalytic structure,
consequently weakening the structure, and the silica vapor
will pass through the reactor into the heat exchanger equipment, condense there, and foul the heat exchanger tubes.

Impregnated Primary Reforming Catalysts

Impregnated reforming catalysts are quite simple to make.
First the proper support material must be selected, this is usually alpha-alumina of the preferred porosity and pore size. The
porosity generally is on the order of 20–30%, and the pore size
is very large in comparison with typical catalysts, being in the
range of fractions of a micrometer rather than the angstroms
typical of catalytic structures. The Carborundum Division of
United Catalysts provides such a support, as does Norton
Company of Akron, Ohio. It should again be pointed out that
silica should not be a part of the composition because of the
sublimination of the silica under the operating conditions for
the steam hydrocarbon reformer reaction with the resultant
problem previously mentioned of weakening of the structure
and fouling of the heat exchanger downstream from the reformer.

Step 1. Alpha alumina in the form of cylinders, rings, or spheres
is charged to a pill-coating device (Fig. 37).
Step 2. The support material is placed in the pilling coating
device, and nickel nitrate solution having a concentration
of approximately 15% nickel nitrate is next sprayed onto

it in sufficient quantity to provide a final quantity of 15% Ni. The nickel nitrate is dried into and on the exterior of the ceramic particles. It should be noted at this time that in addition to the nickel nitrate, chromium nitrate, magnesium nitrate, or aluminum nitrate, for example, can be added in small quantities (3–15% as oxide of the Ni content) to act as a stabilizer for the nickel, which otherwise tends to agglomerate quite severely.

Step 3. The impregnated ceramic particles are now calcined at approximately 900°C to convert the nitrates to oxides and to a certain degree react the nickel oxide with the alpha-alumina or any other stabilizers already coimpregnated onto the alumina support. The catalyst is now ready for use.

It is possible that under some circumstances the quantity of nickel should be greater than indicated in the foregoing instructions. To attain this higher Ni level, it is preferable not to increase the concentration of the nickel nitrate but to impregnate twice, with the second impregnation being performed identically to the first. It is also possible that for the secondary reformer the quantity of nickel should be less than that stipulated in step 2 above; for example, 6–8% nickel is adequate for the secondary reformer. For the catalyst in the secondary reformer, which operates at 1100–1200°C, it is essential that the catalyst be stabilized to prevent agglomeration as previously discussed. It is also essential that no sublimable material such as SiO_2 be present, particularly in the support.

Secondary Reformer Catalyst

This subject was discussed under the final paragraph of the preceding instructions.

Naphtha Reforming Catalyst

Steam reforming of naphtha frequently has as its objective the conversion of the naphtha to CO and H_2, which is the same as with lighter hydrocarbon fractions; however, there may be a difference in that much of the naphtha is converted to methane. This is desirable for the manufacture of a high-methane gas that is to be used as a substitute for natural gas. In fact, the CO and H_2 produced may methanate to CH_4 so that the total methane is above 90% and a high-Btu gas is produced. To achieve this SNG objective, a different type of reformer catalyst is produced. The procedure is as follows:

Step 1. A quantity of alpha-alumina pellets or rings (ca. ⅝ × ⅝ in. to ⅛ × ⅛ in., 20–25% porosity) are placed in a pill coating device (Fig. 37).

Step 2. Sufficient aqueous solution containing 15% nickel nitrate, 3% KNO_3, and 3% $Mn(NO_3)_2$ is added to deposit on the support approximately 8% NiO and corresponding quantities of KOH (K_2O) and MnO_x after decomposing the nitrate.

Step 3. The coated alumina is next calcined at 450°C for 2 hr. Means must be provided for the safe disposal of the evolved NO_x fumes.

The catalyst is now ready to use for naphtha reforming at 500–800°C. (The lower temperatures favor CH_4 production.) Instead of 3% KNO_3, one can increase the quantity to 4% or reduce it to 2%. (One should bear in mind that K_2O sublimes at temperatures above 700°C.) The manganese can be omitted entirely, or it can be increased to 5%. The potassium salt can be the hydroxide or one of the carbonates. Except for sulfate, phosphate, or chloride, another manganese salt can also be substituted for the Mn $(NO_3)_2$. The purpose of the potassium and manganese oxides is to minimize carbon formation or buildup.

REUSE OR RECOVERY OF USED OR SPENT CATALYSTS FOR SYNTHESIS GAS PROCESSING

Zinc Oxide Sulfur Removal Catalyst Adsorbent

Zinc oxide as a catalyst adsorbent is in reality a chemical reagent that eventually is "spent" because of the reaction

$$\left.\begin{array}{l} H_2S \\ COS \\ CS_2 \\ R\text{-}SH \end{array}\right\} + ZnO \rightarrow ZnS + H_2O \text{ or } CO_2$$

The zinc sulfide can be handled in either of two ways, with the least costly and least troublesome being to recycle it to an ore processor who would mix it in with sphalerite ore (ZnS) and roast it at elevated temperatures to form ZnO and SO_2. The ZnO could be recycled for sulfur adsorption, and the ore processor would probably produce sulfuric acid from the SO_2.

The original vendor of the ZnO adsorbent may buy back the used ZnS and roast it at relatively low temperatures, 400–550°C, to convert the ZnS to $ZnSO_4$. The adsorbent, which would be $ZnSO_4$ plus unsulfided ZnO, would be dissolved in H_2SO_4 and then precipitated using an alkali carbonate. This would then be filtered, dried, and calcined at 350°C to produce a highly reactive form of ZnO. DuPont practiced this process with a ZnO-CuO CO-shift catalyst.

CuO or Fe_2O_3 on Activated Carbon as Sulfur Scavengers for Synthesis Gas

Copper and iron oxides on activated carbon are best handled by burning off the carbon and collecting the ash (Fe_2O_3 or CuO) for metals recovery. These constitute such small quantities that jobbers buy the ash and accumulate it until they have sufficient to interest a smelter in its use as a rich ore.

Steam Hydrocarbon Reforming Catalyst: Primary and Secondary Reforming

Steam hydrocarbon reforming catalysts are of significant value because of their nickel content, which varies from about 4% to as high as 50%; alumina comprises the rest of the catalyst. These catalysts can be thought of as a rich ore of nickel, but a smelter is interested only in large volumes, so the jobber is important here also. The alumina or silica-alumina residue becomes a part of the slag that smelters have handled in an environment-friendly manner for many years as road or railroad ballast for similar applications.

Naphtha Reforming Catalyst

Naphtha reforming catalyst is moderated by the addition of an alkali such as potassium carbonate to decrease its tendency to form elemental carbon. The standard unmoderated catalyst cracks the naphtha to carbon. For the recovery of the nickel, the nickel is dissolved in a mineral acid and reprecipitated as the carbonate. The nickel carbonate (or nickel oxide) is washed or if necessary ion-exchanged to reduce the alkali content to less than 50 ppm in the finished nickel oxide.

HIGH-TEMPERATURE CO SHIFT

"CO shift" is a designation generally accepted for the conversion of carbon monoxide in the presence of steam to carbon dioxide and hydrogen. There are three basic processes by which this is effected. One is the high-temperature shift, which operates between approximately 350 and 475°C, and the second is the low-temperature shift, which operates between about 220 and 350°C. The third is an intermediate temperature range using copper oxide promoted composition very similar to the high-temperature shift. The CO shift to hydrogen and carbon dioxide is favored by low temperature, so the low-temperature shift catalyst is usually used in the second stage of the shift operation. The high-temperature shift catalyst is generally an iron oxide–chrome oxide mixture, whereas the low-temperature shift is a zinc oxide–copper oxide–aluminum oxide mixture.

Preparation of high-temperature shift catalyst, as has previously been described in several cases, can be by either a rather refined technique or a rather crude technique. The catalyst prepared by the crude technique is somewhat inferior catalytically in most cases, but it is adequate if the catalyst

is followed by the low-temperature shift. The cruder preparative procedure will be described first.

Step 1. A finely divided iron oxide such as pigment grade material is charged to a kneader such as that depicted in Fig.
27. Then sufficient chromium trioxide (CrO_3) is charged
to the kneader to produce a chromium content relative to
the iron oxide of between 6 and 12%. The chromium oxide
imparts a minor degree of sulfur tolerance to the iron oxide,
which, of course, under operating conditions is a mixture
of iron oxide and elemental iron.

Step 2. Water is next added in sufficient quantity to produce
a paste that is not quite as thick as putty but thicker than
a heavy grease. The kneading is continued for approximately 2 hr, with the temperature maintained at 30–45 °C.
It may be necessary to add water from time to time due
to evaporation from the kneader.

Step 3. The kneaded paste is dried and then calcined at 400 °C
for 3 hr after the temperature reaches 400 °C. Equipment
is shown in Figs. 22, 23, and 26.

Step 4. The calcined catalyst is granulated to 100% through
a 10 mesh screen in equipment such as that shown in Fig.
29 and is then mixed with 1% graphite in a ribbon mixer
(Fig. 31B).

Step 5. The powder mixed with graphite is now pilled in a machine similar to that shown in Fig. 32.

Instead of using the kneader as a means whereby the oxides
are thoroughly mixed together, one can use a ball mill (a pug
mill can also be used), but care must be taken to ensure that
no gases are evolved during the milling, otherwise a rupture
of the ball mill can occur. Furthermore, the catalyst can be
extruded by converting the calcined material to a wet paste,
which then can be extruded (Fig. 28) and dried. The extruded
catalyst may suffer from inadequate strength, and as a consequence it is recommended that pilled catalysts can be used

as a standard material with the possibility that specialized extruded material will attain adequate strength.

A more sophisticated method of catalyst preparation is by coprecipitation of the chromium and the iron:

Step 1. Place in an agitated tank (Fig. 9) a 1 m solution of ferrous sulfate. Then dissolve in the solution chromic acid anhydride (CrO_3) equal to between 4 and 12% chromium oxide (Cr_2O_3) on the basis of the iron oxide eventually derived from the ferrous sulfate.

Step 2. Add ammonium hydroxide (Fig. 9) to this solution until a pH of 7.0 ± 0.2 has been attained. The solution should be maintained at 50°C during the entire period of precipitation and subsequent digestion for 1 hr after the precipitation is completed.

Step 3. Filter the slurry in equipment such as that depicted in Figs. 12–15. Wash it on the filter or by reslurrying until the sulfate content of a dried portion of the filter cake is equivalent to less than 0.5% sulfur.

Step 4. Dry the wet filter cake and then calcine it (Figs. 22–24) at 350°C for 2 hr after the catalyst reaches this temperature.

Step 5. The catalyst may then be densified by kneading it with water in equipment such as that depicted in Fig. 27. The wet paste is dried and on drying is granulated to 100% through a 10 mesh screen (Fig. 29).

Step 6. The screened powder is mixed with 1% graphite (Fig. 30 or 31) and is pilled in a machine similar to that depicted in Fig. 32. The catalyst is now ready for use.

PROMOTED HIGH-TEMPERATURE CO-SHIFT CATALYST

This catalyst, which is a relatively recent development, can be used in place of some or all of the high-temperature CO-shift catalyst discussed in the preceding section. This catalyst is promoted with 0.2–10.0% copper oxide added to the iron-chrome catalyst, which is made by a procedure that itself produces a more active catalyst than the less costly iron oxide-chromia acid anhydride process. The resultant promoted catalyst has a light-off about 30 °C lower than the older high-temperature CO-shift catalyst. A complete process description follows.

Step 1. Make up an aqueous solution comprising a 1 M solution of ferrous sulfate.

Step 2. Add sufficient sodium dichromate to the ferrous sulfate solution of step 1 to produce a Fe/Cr ratio of 100 : 6 to 100 : 10. The higher figure is preferred because it imparts the greatest amount of tolerance to sulfur and halides.

Step 3. With the solution of step 2 at 38 ± 2 °C and being well agitated, add sufficient aqueous 10% solution of sodium hydroxide to bring the pH to 6.8 ± 0.2.

Step 4. Agitate the slurry resulting from step 3 at 38 ± 2 °C for 1 hr, wash it by decantation to free the precipitate of sodium salt, and then filter and wash it additionally on the filter.

Step 5. Dry the filter cake for 24 hr at 125–150 °C.

Step 6. Calcine the dried filter cake at 350 °C for 3 hr after reaching this temperature.

Step 7. Knead the paste to densify it to facilitate the subsequent pilling operation.

Step 8. Dry the kneaded paste at 120–150 °C.

Step 9. After the kneaded paste has cooled, granulate to 100% through a 10 mesh screen in equipment such as that shown in Fig. 29.

Step 10. Mix the powder of step 9 with 1% powdered graphite using the equipment depicted in Fig. 30 or 31.

Step 11. Pill the catalyst to the size and density desired.

Promoting the Catalyst with Copper

Step 12. Prepare and filter cake of copper carbonate by making a 10% solution of copper nitrate at 35 °C and precipitating the copper as carbonate using a 10% solution of sodium bicarbonate. Filter and wash the copper carbonate precipitated to remove the sodium nitrate salt.

Step 13. Assay the filter cake to determine the copper content of the wet filter cake.

Step 14. Weigh the quantity of filter cake necessary to derive the quantity of copper needed to add to a given quantity of catalyst of step 11 (0.2–10%, ordinarily 2.5%).

Step 15. Make a solution of ammine carbonate and ammonia having a 7% content of ammine carbonate and 7% solution of NH_4OH.

Step 16. Put the appropriate quantity of pelleted catalyst into a pill-coating device such as depicted in Fig. 37A–C or D. Add all or a portion of the solution of step 11, as appropriate, rotate the coating pan, and dry the copper ammine carbonate onto the pills.

Step 17. If necessary, heat the coated pills to 300°C for 1 hr to decompose the copper ammine carbonate to copper oxide.

Step 18. The copper oxide-impregnated pills are now ready to use for CO shift.

Note: As an alternative method of adding the promoter, the copper ammine carbonate solution can be added to the kneading operation of step 7. Steps 8–11 are performed, but steps 12–16 are not. The calcining operation of step 17 is performed, after which the catalyst is stored for use as stipulated in step 18.

As stated above, the lower temperature CO-shift catalyst is a mixture of copper oxide, zinc oxide, and aluminum oxide. These, for best performance, are prepared in such a way as to avoid contamination from sodium and sulfate. This means either that the catalyst must be prepared from nitrates and precipitated with ammonium carbonate or, if prepared from sulfates and precipitated with sodium carbonate, for example, the catalyst must be carefully washed to remove the sodium and sulfate to levels of less than 150 ppm each. The procedure in detail is as follows:

Step 1. A solution is prepared in a tank (Fig. 9) having 1 M concentration and comprising a Cu/Zn ratio of 1 : 2. These metals are present as the nitrates. Sufficient aluminum nitrate is also added to the solution, after the catalyst has been calcined to oxides, for the aluminum oxide level to be equal to 12% of the total.

Step 2. Ammonium bicarbonate, either as a solid or as a 10% solution, is added to the copper-zinc-aluminum nitrates solution, which is being maintained at 30°C.

Step 3. Continue the addition of the precipitant until a pH of 6.8–7.2 has been reached. Allow the slurry to agitate and digest for 1 hr, and then filter it (Figs. 12–15, 21).

Step 4. Filter the catalyst, washing it on the filter sufficiently to remove essentially all of the occluded ammonium nitrate salts.

Step 5. Dry the catalyst in equipment such as that shown in Fig. 18 or 20, and then calcine it at 400°C in equipment similar to that shown in Figs. 22, 25, or 26.

Step 6. It is possible that the calcined filter cake will be sufficiently hard and dense that upon granulation through a 10 mesh screen and mixing with 1% graphite in a ribbon mixer, it will produce a powder density that will pill satisfactorily. (Powder having an apparent density of 0.55–0.80 g/mL should pill well.) The calcining is performed in equipment such as that illustrated in Figs. 22, 25, or 26, whereas the granulation, mixing with graphite, and pilling are each performed in equipment depicted in Figs. 29, 30, 31, or 32. After pilling, the catalyst is finished and ready for use.

The ratio of zinc to copper can be varied according to the preference of the plant operator. The 2 Zn/1 Cu ratio given above is the typical ratio, but higher or lower proportions of copper have been explored and are preferred by some. A cruder method of preparation of this catalyst has also been examined by some people who seem to be satisfied with it. This simply involves mixing copper oxide, zinc oxide, and aluminum oxide or alumina hydrate in the preferred proportion in a kneader, ball mill, or pug mill. It is also possible to combine the two methods of preparation by attempting to hydrolyze the ingredients into the most active form while maintaining the low-cost features of the cruder mixing method. Hydrolysis is effected by a hydrothermal treatment at or near 100°C with or without CO_2 bubbling through the hot slurry.

The last of the operations in gas synthesis is methanation, which is a process whereby the last traces of carbon monoxide are converted to methane. Methane, in contrast to CO, is non-poisonous in most applications of hydrogen.

METHANATION CATALYST

Methanation catalyst usually comprises nickel or ruthenium on a support. In preparing the nickel-on-support methanation catalysts, which will be described first, two general types of procedures can be followed. As we have indicated in other preparations, one is coprecipitation and the second is impregnation. Each procedure and each catalyst has its advocates, generally based upon preference for adequate performance at low cost or better performance at a higher price.

Step 1. A 1 M solution of nickel nitrate is prepared in a tank such as that shown in Fig. 9. Also slurried in that tank is sufficient alumina hydrate to give a content of aluminum oxide in the final catalyst equal to approximately 25% of the nickel oxide present. The alumina can be varied from 10 to 80% as preferred by individual users of the catalyst.

Step 2. Heat the solution to 75°C and add either solid or a 10% solution of ammonium bicarbonate until the pH reaches 6.8–7.2.

Step 3. Let the slurry agitate for an additional 1 hr, and then immediately begin to filter it through equipment of the type shown in Figs. 12, 13, 14, 15, or 19. Dry the filter cake in equipment such as that shown in Figs. 17, 18, or

20 and calcine it for 3 hr at 400°C after the catalyst has reached this temperature in equipment of the type shown in Figs. 22, 24, and 26.

Step 4. The calcined catalyst must be densified by kneading with water to form a paste, not as stiff as putty but stiffer than a heavy grease, in kneading equipment such as that shown in Fig. 27. Dry densified powder and then screen it to 100% through a 10 mesh screen in equipment such as that depicted in Fig. 29.

Step 5. Mix granulated powder with 1% graphite in a ribbon mixer such as that shown in Fig. 30 or 31 and then pellet it to form hard, low-density catalyst in pilling equipment such as that shown in Fig. 32.

In addition to the alumina called for in step 1, or substituting for some or all of it, one can use chromium, magnesium, thorium, zirconium, lanthanum, or other refractory oxides.

This completes the description of the more sophisticated preparation techniques. What follows is the procedure for the supported methanation catalysts, a simpler and sometimes quite adequate preparation procedure.

Step 1. A high surface area alumina (usually gamma-alumina) support with a surface area in the range of 150–300 m²/g and of the granule size required for the process is selected. This size may be as large as 2–4 mesh granules to as small as 16–20 mesh granules or ⅛–¼-in. spheres, or extrudates or pellets in a corresponding range of sizes. The selected weight of one of these is charged to a pill-coating device such as shown in Fig. 37.

Step 2. With the pill-coating device rotating, a solution of nickel nitrate or nickel ammine nitrate is added in sufficient quantity to provide an 8–15% quantity of nickel relative to the alumina of the alumina support. The nickel salt is dried on the support by using a jacketed heater on the pill-coating device.

Step 3. The coated pellets are calcined in equipment such as that shown in Fig. 22 or 26 at 400°C for a period of 1–4 hr to convert the salt to nickel oxide. After it has been reduced, generally in the reactor itself, the catalyst is ready to use in the methanation process.

RUTHENIUM ON ALUMINA METHANATION CATALYST

Ruthenium is a highly active methanation catalyst and is especially useful at low temperatures and pressures. However, it has a well-known reputation for the Fischer-Tropsch reaction, and if it is operated under other than optimum conditions (too high pressure) it will synthesize wax or high-boiling organics on the surface that will not be desorbed and will poison the catalyst. The procedure for its preparation is as follows.

Step 1. Activated alumina having a surface area of 150–300 m²/g and in the form of granules (from 2–4 mesh to as small as 18–20 mesh), ⅛–⅜-in.-diameter spheres, or pellets or extrudates are charged to a pill-coating device (Fig. 37).

Step 2. Sufficient 5% sodium bicarbonate solution is added to the alumina in the pill-coating device to just moisten the alumina without excessive liquid being present on the walls. This is accomplished by making up a solution of exactly the volume as the total pore volume of the support in question.

Step 3. The alumina, still moist with sodium bicarbonate solution, is now dried either by removing it from the pill-coating device or by turning on the jacket heat of the pill-coating device and rotating until dry.

Step 4. The dry alumina impregnated with sodium bicarbonate is returned to the pill-coating device and coated with ruthenium chloride. This is accomplished by making a solution of ruthenium chloride of essentially the same volume as that which was used for the sodium bicarbonate solution mentioned in step 2. The ruthenium content is adjusted to provide 0.05–0.3% Ru on the support.

Step 5. The ruthenium chloride solution is sprayed onto the alumina support, and, because of the presence of the sodium bicarbonate, the ruthenium is deposited as a thin coating on the periphery of the support material (Fig. 37).

Step 6. The rotation of the pill-coating machine is continued for 15 min to ensure complete precipitation of the ruthenium, and then distilled or demineralized water is added to the pill-coating machine and is allowed to overflow until such time as all the sodium that is extractable is removed from the pellets.

Step 7. The washed catalyst is removed from the pill-coating device and is dried at 150°C. After drying, the catalyst is ready for use in the methanation reaction.

NICKEL CHROMITE TYPE METHANATION CATALYST

Nickel chromite, when reduced, becomes a very effective, relatively low temperature, and long-lived CO methanation catalyst. Methane leakings at 230°C, 25,000 space velocity, and 400 psi can be below 5 ppm with a 6000-ppm input. The catalyst is sulfur- and halide-resistant.

Chromium, with a valence of +6, is a particularly toxic chemical, so processing and disposal must be carefully planned and executed so that dust and personal contact is rigidly minimized and preferably avoided. Two methods of preparation follow, the second of which avoids waste materials that must be sewered or disposed of.

Coprecipitation Procedure

Step 1. Make an aqueous solution having an 0.5 M content of nickel nitrate hexahydrate and 0.55 M content of chromic acid or ammonium chromate at 40 ± 2°C.

Step 2. While the solution of step 1 is being agitated, add 28% ammonium hydroxide to reach a pH of 6.9 ± 0.1. This causes the precipitation of a basic nickel ammonium chromate.

Step 3. Filter the slurry and wash it on the filter. The filtrate must be stored for chromium recovery or for use as the solvent in the next preparation. All wash waters must be processed for Cr^{6+} removal.

Step 4. Dry the filter cake at 125–150°C and then calcine it at 450°C for 2 hr after reaching this temperature.

Step 5. Knead the calcined powder to densify it.

Step 6. Dry the kneaded paste and then granulate it by pass-it through a 10 mesh screen.

Step 7. Mix the dried granulated paste with 1% graphite, then pill it in a pelleting press.

Step 8. The pilled catalyst is stored until used. The catalyst, when charged to the unit, must be carefully reduced before use.

Zero Discharge Method of Preparation of Nickel Chromite

The following procedure is a scheme that produces a fairly active catalyst but has as its important advantage the fact that there is no liquid or solid by-product presenting a disposal problem.

Step 1. A mixture of 3 parts by weight of low-density nickel oxide powder (this is an article of commerce) and 1 part of dimineralized water is placed into a kneader or a heavy-duty mixer. To this mixture is added slowly 3.5 parts by weight of ammonium chromate. The mix is kneaded to produce a reaction product of $Ni(NH_3)_xCrO_4$. This, if the ingredients are correctly selected, will be a greaselike paste. If desired, one part of finely divided alumina hydrate can be added to give a nickel oxide chrome alumina mixture, with some inter-reactions producing spinels and other reaction products.

Step 2. The paste is dried, then calcined at 400°C for 2 hr after reaching 400°C.

Step 3. The calcined paste is granulated to 100% through a 10 mesh screen.

Step 4. The granulated powder is mixed with 1% graphite and is pilled.

Step 5. The pilled catalyst is now ready for charging to the methanation unit and careful reduction for use.

METHOD FOR REUSE, REGENERATION, AND METALS RECOVERY FOR METHANATION CATALYSTS

The simplest procedure for recovering metals from methanation catalysts is a regeneration in place by either a gas or liquid treatment. The gas treatment may be a careful reoxidation and simple rereduction. An alternative procedure is reoxidation followed by removal from the unit, further reoxidation externally, and then rescreening and recharging. This method is suitable when combustible or volatile materials must be removed.

A third scheme involves the foregoing plus extraction with acid or hydrothermal treatment as an aqueous slurry at 100 ± 5°C.

A fourth is reimpregnation with one or more of the active ingredients and calcining for reduction to the active form.

The fifth option, a more severe treatment, involves reoxidation, removal, milling with a liquid, reslurrying, and precipitation of one or more of the ingredients on the finely divided used materials. The product is then processed and employed as new material. This procedure is made a bit more effective and costly by the further two steps of dissolving and and reprecipitating one or more of the ingredients. With nickel chrome or nickel alumina, the obvious ingredient would be the nickel.

If all of the foregoing fail to produce a commercially satisfactory product, there are a number of companies (see Appendix) that purchase scrap catalysts for reprocessing much like
an ore. They are necessarily wary of spent material that may
be contaminated with toxics, and it is the responsibility of the
catalyst user to remove toxic ingredients before sending spent
catalyst to the recovery agency.

In the case of ruthenium, reference should be made to the
procedures described on pp. 254–257 and in Chapter 30.

16

Ammonia Synthesis Catalysts

Ammonia synthesis catalysts at this time are very similar in composition and form to that first used by Haber in the early 1900s. It was a natural magnetite, which fortuitously contained the promoters of almost the identical type and quantity presently used. This catalyst was used by Haber in his early experiments and served quite satisfactorily. Today, much of this commercial catalyst is still made from natural ore, but some of it is made by oxidation of iron bars or specially selected scrap iron. When using natural ore or scrap iron, it is essential that the level of copper be held to the minimum (less than 100 ppm) because of its severe poisoning of the ammonia synthesis catalyst.

Step 1. Crushed magnetite (Fe_3O_4) in the form of 8 mesh material is carefully mixed with promoters. These promoters, based on the weight of the magnetite, are approximately

1% potassium as carbonate, 1–3% calcium carbonate, 2–5% aluminum oxide, and occasionally 0.5–3.0% chromium oxide. The chromium oxide may be added as potassium chromate or dichromate.

Step 2. The magnetite and the promoters are all carefully mixed together to form a uniform mixture.

Step 3. The mixture is now placed in a large pan into which two or three water-cooled electrodes can be immersed.

Step 4. An arc is struck from one electrode by a hand-held electrode, which permits the formation of a small amount of fused iron oxide and promoter, and this movable electrode is finally moved over to an opposite electrode. This is repeated for all three electrodes, if three are present (three phase).

Step 5. The fusion furnace represented in the block flow sheet of Fig. 6 has a controlled ampere-voltage system that permits the voltage to drop as the amperage increases, and eventually a large ingot of molten magnetite with promoter is formed.

Step 6. When the ingot of the typical size possible in the fusion furnace has formed, the electric energy is turned off and the water-cooled electrodes are allowed to remain in place.

Step 7. After several hours of cooling, the electrodes are removed from the melt. By that time the ingot has hardened and can be carefully lifted from the unmolten magnetite mix, which forms a crucible for the molten iron oxide promoters. The ingot is removed to a location where it can be allowed to cool, which usually requires at least 24 hr; for safety reasons, 48 hr is preferable.

Step 8. The cooled ingot is crushed using the jaw or cone crusher (Figs. 34 and 35).

Step 9. The crushed material is now screened using the equipment depicted in Fig. 36. The proper mesh size is selected from experience; for conventional operations it is usually in the range of 2–4 or 2–3 mesh. The catalyst is now ready for use after careful reduction.

The magnetite called for in step 1 can be purchased as magnetite ore from certain locations. Sweden has an excellent supply, and much of the magnetite used in the United States comes from that source. A synthetic magnetite can be made by oxidizing specifiable iron bars, rods, or other convenient shapes. The bar is ignited in a flow of oxygen, and the iron oxide formed drops like melted candle wax into a receiver in the bottom part of the combustion chamber. This method of producing magnetite gives a relatively pure magnetite, except that the oxygen level is slightly above the Fe_3O_4 range. However, during the subsequent fusion called for above, the excess oxygen is liberated and a magnetite composition results.

Although this catalyst is identified as an ammonia synthesis catalyst, it also was one of the leading candidates in early German investigation of the Fischer-Tropsch synthesis.

SINTERED AMMONIA SYNTHESIS CATALYSTS

The fusion operation is one that is contrary to the basic concepts of catalyst preparation and use. Fusion is more severe than sintering, which is usually considered one of the major causes of catalyst failure during use. Some very effective catalysts have, however, been made by sintering, which can be initiated at a temperature about 60% below the fusion point (500–600°C). With sintering conditions, much porosity and small crystallites remain. Some work has been reported on a method for preparing a çoprecipitated into a mixture of iron oxide, alumina, and magnesia, subsequently promoted with potassium carbonate or chromate, and finally sintered at 600–900°C.

Despite the fact that this catalyst was reported to be active at temperatures 50–125°C below the 425–500°C range for the fused catalyst, there is no report that it was ever commercialized.

If there really was this temperature advantage, the catalyst could offer major economic advantages in reduced pressure of operation, reduction or elimination of refrigeration of reactor effluent, or reduced recycling of unreacted gases.

RECOVERY OF SPENT CATALYST

Spent ammonia synthesis catalyst is pyrophoric when it is removed from the unit unless it has been reoxidized before discharge from the unit. Even though it is "reoxidized," there may be pockets or lumps of catalyst that are still in a reduced condition. Extreme caution is required.

The spent, completely reoxidized catalyst is a good raw material for a smelter, but smelters are interested only in large quantities. Some of the jobbers listed in the Appendix purchase smaller amounts of spent catalyst and sell it when sufficient has accumulated to interest the smelter.

17

Methanol Synthesis Catalysts

There are two basic types of methanol synthesis catalysts. One is a high-temperature, high-pressure catalyst and the other is a low-temperature, low-pressure catalyst. These designations are arbitrary and are not accurate as to the service in which each will perform. The high-temperature catalyst can be made to function at relatively low temperatures but not as low as that identified as a low-temperature catalyst. The low-temperature catalyst, by suitable modifications, can be made to operate at high temperatures and pressures. There is a greater tendency with the low-temperature catalyst to produce by-products when operated at high temperatures and pressures, but this also can be adjusted by modifying the composition. The first preparation to be described will be that of the high-pressure, high-temperature catalyst.

HIGH-TEMPERATURE METHANOL
SYNTHESIS CATALYST

Step 1. Place a high purity, finely divided zinc oxide powder
in a kneader of the type depicted in Fig. 27. Next, add sufficient chromic acid (CrO_3) anhydride to the zinc oxide in
the kneader to be equivalent to the stoichiometric amount
necessary to convert between 30 and 60% of the zinc oxide
to zinc chromate and leave a residue of unreacted zinc
oxide.

Step 2. Next, add to the kneader sufficient distilled water to
produce a heavy paste, not unlike putty but slightly less
viscous. Heat will be generated during the course of the
addition of the water, and the kneader should be jacketed
with cooling water to dissipate the heat and maintain the
temperature below 50–55°C in the kneader. Continue the
kneading until complete reaction has taken place, which
usually is a matter of 45–60 min.

Safety Note: The vapor rising from the kneader may contain small quantities of vaporized chromium compounds. Extreme care must be exercised to avoid inhaling these compounds, because they cause serious nasal and bronchial problems. Ventilation of the kneader must be very thorough.

Step 3. After kneading has achieved the desired reaction and
homogenizing of the paste, remove and dry the paste. After
drying, calcine the catalyst for 2 hr at 400°C after the catalyst has reached this temperature (Figs. 22 or 23).

Safety Note: Ventilation and safety clothing meeting the
most severe requirements must be used because of the Cr^{6+}
content and resulting vapor and dust.

Step 4. Granulate the catalyst to 100% through a 10 mesh
screen. Then mix it with 1% graphite and finally pill it in
a pilling machine of the type shown in Fig. 32.

Step 5. Carefully calcine the catalyst at 475°C to cause partial oxidation of the graphite, thus opening some of the pores of the catalytic material.

Step 6. Reduce the catalyst either externally to the reactor or, having charged it to the reactor, by passing hydrogen in a concentration of 0.5-1% in nitrogen until no further heat rise is obtained or there is no evidence of reduction, which is generally indicated by the absence of moisture appearing in the exit gas. As long as reduction is taking place, moisture will appear in the exit gas.

Step 7. If the catalyst is reduced external to the operating unit, cool it, place it in storage vessels, and cover it with an inert gas. Then store it or transport it to the location where it will be used.

LOW-TEMPERATURE, LOW-PRESSURE METHANOL SYNTHESIS CATALYST

This is the well-known catalyst used generally at the present time at pressures below 2000 psi; it contains various ratios of copper oxide, zinc oxide, and aluminum oxide, the latter as a stabilizer. What is described in the following is the type generally accepted as a preferred low-temperature methanol synthesis catalyst. Note the similarity between this catalyst and the one produced for low-temperature CO-shift operations described in Chapter 15.

Step 1. In a suitable vessel (Fig. 9), make a 1 M solution comprising zinc and copper nitrates with a ratio of 2 Cu to 1 Zn, atomic basis. Also dissolve sufficient aluminum nitrate in the solution to equal 10-18% aluminum oxide in the final product after it has been converted to the respective oxides.

Step 2. With the copper, zinc, and aluminum nitrate solution at 50°C, add sufficient 10% sodium bicarbonate solution to the zinc-copper solution to increase the pH to 7-7.3 range.

Step 3. Allow the slurry thus prepared to digest at 50°C for an additional 60 min in order to effect complete reaction and precipitation.

Step 4. Filter the slurry in any of the types of equipment shown in Figs. 12–15.

Step 5. When the sodium ion has been carefully removed from the filter cake by washing on the filter, dry the cake and then calcine it at 400°C for a period of 2 hr, after the temperature reaches 400°C, in equipment as shown in Figs. 22, 23, or 26.

Step 6. Granulate the calcined catalyst to 100% through a 10 mesh screen in equipment as shown in Fig. 29. Note: Analysis of the calcined catalyst should be made for sodium in the calcined powder. If the content is above 100 ppm, the powder must be slurried in a solution of 0.3% NH_4HCO_3 on a basis of 1:20 powder weight to solution weight. Wash (ion exchange) repeatedly until a Na content of < 100 ppm is reached. Then dry, granulate, and proceed with step 7. After granulation, mix the powder with 1% graphite in a ribbon mixer as shown in Fig. 31.

Step 7. Pill or extrude the catalyst in equipment as shown in Figs. 28 or 32. After forming into pellets or dried extrudates, the catalyst is ready for use. It must, however, be reduced in situ in the methanol synthesis reactor.

In both the low-temperature and high-temperature catalysts, the ratio of copper to zinc or of zinc to chromium can be very sharply modified. Furthermore, copper as copper oxide can be added to the zinc chromite composition to give a copper-zinc-chromium so-called high-temperature, high-pressure catalyst, which under the influence of the added copper will now be suitable for operation at lower temperatures and pressures. These modifications can be made at the discretion of the researcher in cooperation with the user of the catalyst. It should be borne in mind that if copper is added to a high-temperature, high-pressure catalyst and the temperature is not reduced, the catalyst under these conditions may very well produce excessive quantities of products such as methane or dimethyl ether. As previously stated, sodium must be washed from the catalyst so the final catalyst has less than 100 ppm sodium; otherwise a relatively inactive catalyst will be produced. A further note is that both types of catalysts must be reduced before they will synthesize methanol, and the reduction must be conducted in a very careful manner so as to avoid overheating, which is usually accomplished by having a low concentration of hydrogen such as 0.5–1% in nitrogen as the reducing gas.

RECOVERY OF VALUES FROM BOTH HIGH-TEMPERATURE, HIGH-PRESSURE AND LOW-TEMPERATURE METHANOL SYNTHESIS CATALYSTS

The high-temperature, high-pressure methanol synthesis catalyst has a composition of $Cr_2O_3 \cdot CuO \cdot ZnO$ or Cr_2O_3, ZnO. Both contain chromium and for this reason cannot be used for landfill because the chromium oxidizes in the landfill environment to water-soluble Cr^{6+}. This form of chromium is toxic, and if it enters the aquifer it can be harmful to beneficial organisms. Consequently this catalyst must be disposed of by processors or jobbers listed in the Appendix.

Low-temperature methanol synthesis catalyst has a composition of $CuO \cdot ZnO \cdot Al_2O_3$ in various proportions. It also must be reduced before use and usually is dumped from a reactor in at least partially reduced condition. Consequently it may actually produce a flame and heat from adsorbed organic materials and catalyst reoxidation when exposed to air. After reoxidation the catalyst can be pulverized and redissolved in a dilute nitric or sulfuric acid solution to produce the desired salts. If the alumina at this stage is equivalent to the original alumina or alumina salt, then the solution or solution-slurry (insoluble Al_3O_3) is processed by precipitation, washing, filtration, calcining, etc., as for a new catalyst from new salts. The spent catalyst thus is used as a source of ingredients that are processed as in a fresh catalyst preparation.

If the spent low-temperature methanol is contaminated with materials that make the foregoing procedure unwise, then, after careful and complete oxidation, it must be handled by one of the processors or jobbers listed in the Appendix. Protective clothing and breathing equipment must be worn, and any other appropriate safety equipment must be used.

18

Hydrogenation Catalysts

ELEMENTAL NICKEL SUPPORTED ON A FINELY DIVIDED SUPPORT

Nickel on kieselguhr is a common catalyst used in the reduced condition as a powdered nickel-bearing catalyst useful for liquid-phase hydrogenations primarily for hardening or converting vegetable oils to a soft wax, for instance, soft oleomargarine. Raney nickel is also used broadly in this operation as well as others and is described in a later section of this chapter. The procedure for preparing nickel on kieselguhr follows.

Step 1. Place in a precipitation vessel a 1 M solution of nickel nitrate. Then slurry in this solution sufficient finely divided kieselguhr (such as that having the trade names Filtercel or Dicalite) of a quantity sufficient to be equivalent to 1–3 times as much kieselguhr as elemental nickel in the final product.

Step 2. Heat the solution-slurry to 70°C, and then slowly over a period of 2 hr add a 10% sodium bicarbonate solution so that a pH of 7.6 is reached for the nickel carbonate slurry.

Step 3. Allow the slurry to continue agitation after the pH has reached 7.6 for an additional 16 hr of digestion at 70°C. The purpose of this digestion is to get an interaction between silica and the nickel carbonate, which in most cases is beneficial to the activities of the catalysts.

Step 4. After the digestion period, filter the slurry in or on equipment such as that shown in Figs. 12–15.

Step 5. Dry the catalyst and finally granulate it to 100% through an 8 mesh screen in facilities shown in Fig. 29.

Step 6. Reduce the catalyst by passing 100% hydrogen over or through it in a suitable enclosed vessel, usually of a specific design for the operation. For laboratory operations, a glass tube is inserted in a split tube furnace as shown in Fig. 25. Hydrogen of 100% concentration is passed over the catalyst at 500°C until such time as it is completely reduced. Usually this can be accomplished in a matter of 2 hr.

Stabilization of the Catalyst

The catalyst in reduced form is pyrophoric; that is, when exposed to air, it will immediately glow, oxidize, and become catalytically ineffective. To make it possible to use and handle it in air, it is given the following stabilization treatment. After reduction and cooling, the catalyst is transferred into an inert atmosphere in, for example, a glass bottle. A stopper is inserted into the bottle, and inserted in the stopper is a thermometer extending down into the catalyst in the bottom of the bottle. Periodically, air is admitted to the bottle and the temperature immediately rises by as much as 5–15° from an original temperature of approximately 15°C. When the catalyst reaches 35°C after periodic addition of air, the bottle and its contents are cooled by immersion in an ice bath or something similar. The cooling and admission of air are repeated until there is no further temperature rise, which usually requires four or five cycles of admitting air and cooling. The catalyst is now stable to a temperature of 75°C in air.

In commercial operations, the reduction, as previously indicated, requires special equipment, and the stabilization similarly requires special processing equipment. These must be engineered for the specific needs of the process. After reduction, the catalyst is usually placed in a drum, which is later rotated and cooled to 15°C, with air continuously added. The temperature is monitored until there is no further temperature rise.

The nickel-on-kieselguhr catalyst can be promoted and activated by coprecipitating the following materials: colloidal silica from Ludox, preferably as Ludox SM; 5–15% silica addition on the basis of replacement or in addition to the kieselguhr is usually satisfactory. Cerium oxide in similar concentrations from cerium nitrate, aluminum oxide from aluminum nitrate, and zirconium oxide in small concentrations from zirconium nitrate are also effective and can be especially useful in specific cases. Titania also is useful and can be added as Tyzor, a commercial organotitanate. Some research would be necessary to use these promoters in specific applications.

COBALT ON KIESELGUHR

Cobalt has various specific applications, usually in the hydrogenation of nitriles to amines or in other cases where an amine is present as a substituent group. Cobalt on kieselguhr can be made in essentially the same procedure as nickel on kieselguhr by substituting cobalt nitrate for the nickel nitrate stipulated in step 1 of the previous section.

NICKEL ALUMINUM ALLOY (RANEY CATALYST)

Nickel aluminum alloy catalysts, either as powder or in lump form, frequently have characteristics that make them superior for some hydrogenations. In the case of the powder, one of the advantages is that it is extremely active and has a high settling rate so that a reaction can be conducted and the catalyst allowed to settle and then the product can be decanted from the catalyst, which then can be used in the next batch. The procedure for preparing the nickel aluminum alloy follows.

Step 1. Elemental nickel in the form of approximately ¼-in. shot is heated to 900°C in a graphite crucible.

Step 2. Molten aluminum is poured over the nickel shot in such quantity as to give a weight ratio of aluminum to nickel of 1 : 1 if the catalyst is to be in powder form, or 58 : 42 if the alloy is to be used in lump form.

Step 3. The molten alloy is poured into one or more molds for cooling.

Step 4. If the alloy is to be used in lump form, it can be crushed and screened in equipment such as that shown in Figs. 34–36. However, if it is to be used in powder form, it can be milled dry in a ball mill available from suppliers such as Paul O. Abbe.

Step 5. The alloy is activated by exposure to a sodium hydroxide solution at a pH of approximately 14. The temperature is maintained at 35°C or below, and the pH is never allowed to go below 11.5, for if it does, aluminum hydroxide may be precipitated on the exposed nickel and an adverse affect produced.

Reactions: $2 \text{ Al} + 2 \text{ NaOH} + 2 \text{ H}_2\text{O} \longrightarrow \text{Na}_2\text{Al}_2\text{O}_4 + 3\text{H}_2$
$\text{Na}_2\text{Al}_2\text{O}_4 + 4\text{H}_2\text{O} \longrightarrow 2\text{Al(OH)}_3 + 2\text{NaOH}$

If the catalyst is used in lump form, this activation is generally conducted in situ with a continuous flow of sodium hydroxide at pH 14, a temperature below 35°C, and an exit pH above 12.5

Safety Note: Means must be provided for the safe disposal of the H_2.

If the alloy is to be used in the form of surface-activated lumps, the catalyst is melted and cast as 58% aluminum, 42%

nickel alloy ingots. This ratio makes a product in which after crushing and screening the granules are dimensionally stable, whereas the typical 50% Ni, 50% Al alloy will permit the granules on exposure to air to spontaneously disintegrate to a powder. As stated above, the 50% Ni, 50% Al is used in powder form, so the breakdown to powder is not a problem but a desirable normal action.

Another very important factor is that the 42% Ni, 58% Al ratio provides the exact composition to produce the $NiAl_3$ species. However, if the alloy is produced by heating the nickel to 800 ± 50°C and the exothermal effect is generated by pouring molten aluminum over it, the eventual temperature is 1200–1400°C. When the melt is then cold cast, the cold ingots are about 40% $NiAl_3$, 35% Ni_2Al_3, and 25% NiAl. The NiAl is not activated by alkali, and the Ni_2Al_3 produces a low-activity product. If it is desired (is economically feasible), the ingots can be annealed at 800–850°C – for 30 min at 850°C or 1 hr at the lower temperature. When the ingots are cooled, the phases will equilibrate at 70+% $NiAl_3$, with the other phases proportionately reduced.

The resultant carefully (see above) activated lumps are 50–150% more active than lumps that have not been annealed.

COBALT ALUMINUM ALLOY

Cobalt alloy is produced in much the same way as that presented above for nickel. In both cases the activated catalyst must be sequestered from oxygenated materials such as air, water, or methanol that might react with the highly reactive metal, which would be deactivated by this oxidation. Furthermore, hydrogen may be generated, and if the catalyst is kept in a closed vessel a destructive explosion may occur. In some preparations described by others, the temperature is raised to the boiling point. This temperature can be used for the powder; however, we do not recommend that procedure.

RECOVERY OF THE Ni OR Co VALUES FROM
SPENT NICKEL OR COBALT ALUMINUM

Safety Note: The spent finely divided metals are extremely adsorptive and reactive with air (oxygen) and even water or methanol. Because of their adsorptivity they will retain large percentages of contaminants from the reaction in which they were used. Consequently, it should be assumed that they contain toxic material and should be handled accordingly. Furthermore, the finely divided elements are usually pyrophoric and, together with the usual combustible contaminants, present a severe spontaneous fire hazard.

As stated above, the finely divided metal reacts slowly with water, methanol, or other oxygen-containing material to generate hydrogen: $Ni + H_2O \rightarrow NiO + H_2$. This means, of course, that if the spent catalyst is put into a gastight container, the hydrogen generated can produce sufficiently high pressures to cause the container to rupture.

Safety Note: The powdered spent material is extremely reactive with acid and can react violently. If the acid level is allowed to reach a high level (a pH of 2–3) while cold and the temperature rises rapidly, the reaction may become so vigorous that it overflows the vessel. To avoid this, heat the slurry of spent catalyst to 98–100°C, and add the acid a small amount at a time. The reaction will take place and quickly subside because the acid is consumed. Repeat these two steps until no further reaction takes place.

Safety Note: If nitric acid is used to dissolve the spent catalyst, extreme care must be exercised to manage the NO_x that is likely to be produced. If a nitratable organic is present, it is essential to be aware of explosively unstable organonitrates and practice proper precautions. If hydrochloric or sulfuric acid is used to dissolve the spent catalyst, hydrogen will be produced, causing pressure buildup and/or flammability problems.

COPPER CHROMITE SELECTIVE
HYDROGENATION CATALYST

Copper chromite is one of the most effective catalysts for the hydrogenation of esters to produce alcohols without excessive hydrogenation to hydrocarbons. Copper chromite was one of the first spinel-type catalysts produced and has been very useful for the operation in question. The preparative method is as follows:

Step 1. A solution is prepared containing 0.5 M concentration of copper as nitrate and 0.5 M chromic acid as CrO_3.

Step 2. This solution is heated to 30°C and rapidly agitated in a vessel such as that shown in Fig. 9. While at this temperature and being vigorously agitated, anhydrous ammonia is bubbled into the system until a pH of approximately 6.8 is reached. Under these conditions, the copper is precipitated as a complex usually identified as a basic copper ammonium chromate [$Cu(NH_4)_3OH\ CrO_4$].

Safety Note: In handling this chromium-containing catalyst and the chromic acid, extreme care must be exercised and ventilation provided to avoid nasal problems for anyone likely to be exposed.

Step 3. After completion of the precipitation and an additional 1-hr digestion period, the catalyst is filtered in equipment such as that shown in Figs. 12–15.

Step 4. The catalyst is now dried and eventually calcined at 450°C for 2 hr after reaching temperature. Equipment such as that shown in Figs. 22–24 or 26 is used. If the catalyst is to be used as a powder, it is now ready for use after calcining. If, however, it is to be converted to an extrudate or pelleted form, it is processed by densification in equip-

ment such as the kneader shown in Fig. 27; if so desired,
it can then be extruded in equipment such as that shown
in Fig. 28.

Step 5. If desired, the copper chromite can be promoted by
adding the appropriate agent(s) as a solution to the kneader.
Useful promoters (and pelleting aids) are 1–10% $MgCrO_4$,
$CaCrO_4$, or $SrCrO_4$.

Step 6. If it is to be pilled, the kneaded catalyst is dried and
then granulated to 100% through a 10 mesh screen and
finally mixed with 1% graphite in a ribbon mixer such as
that shown in Fig. 31.

Step 7. The powder mixed with graphite is now pilled in equipment such as that shown in Fig. 32.

Step 8. The fixed bed of pellets usually must be reduced, and
this is accomplished at 250–350°C using a reducing gas
comprising 0.5–2% H_2 in N_2 or CO_2. The final stage of reduction is with 100% H_2, but only after no exotherm is
experienced at lower H_2 levels. The reduced catalyst is
pyrophoric and must be carefully reoxidized before discharge.

The foregoing procedure can be modified to include a support by slurrying a support such as kieselguhr or alumina hydrate in the slurry called for in step 1. The procedure can also
be modified to produce an iron chromite, cobalt chromite, zinc
chromite, nickel chromite, or manganese chromite, and other
spinels can be produced by substituting the appropriate ion
for the copper. Mixed spinels such as zinc-copper, zinc-man-
ganese, cadmium-copper, etc., can also be produced in this
manner. Zinc-copper chromites are active methanol synthesis
catalysts. The general procedure given above produces spinels
that have excellent properties for many reactions.

Recovery and Reuse of Copper Chromite Catalysts

If the copper chromite has been used in a fixed-bed operation
as pellets and has become partially deactivated, it must be

reoxidized and the combustibles removed before discharge. The reoxidation consists in adding a gas flow of 0.2–0.5% O_2 in N_2, CO_2, or a mixture thereof until there is no further exotherm and the oxygen level has been increased to 21%.

The reoxidation of the used catalyst may accomplish one or more of the following:

A. Removes carbonaceous species, which may be the sole or principle cause of deactivation.
B. Removes sulfides or halides that contributed to or caused deactivation.
C. The results of items A and B may reactivate the used catalyst so that some or all of it can be reused at least once. Under the most favorable conditions the use–reoxidation–reuse cycle can be repeated for five or more times.

If A and B regenerate only a portion of the bed that can be recycled, usually the upstream portion of the reactor, the unusable portion is usually sold to a jobber for use as an ore supplement for a smelter.

If, on the other hand, the copper chromite or other catalytic spinel is used as a powder in an organic liquid (to hydrogenate coconut oil to straight-chain alcohols), the used catalyst may be heavily contaminated with oil. Before further processing, the oil must be removed, usually by careful oxidation and simultaneous reoxidation of the catalyst. After the reoxidation, (1) the catalyst may be fully reactivated and capable of being reused repeatedly, or (2) the catalyst may be partially reactivated and capable of again being beneficially converted to a powder and added to a new preparation as a "support" during precipitation. Usually the incorporation of only 10% new catalyst during such precipitation produces a product equivalent to completely new catalyst. Advantages of a catalyst of this latter type are that it will have superior filtration properties or it may be dense enough to settle rapidly and permit a process that entails continuous operation with continuous bottom feed and continuous draw-off at the top of the reactor, with the catalyst agitation and settling rate being so designed that the draw-off liquid is free of catalyst.

Pelleted copper chromite is an excellent oxidation catalyst that has low-temperature activity and total oxidation of volatile organic compounds (VOCs). The used pelleted catalyst after reoxidation should be active for VOC oxidation even if it is not active enough to be returned to the original ester hydrogenolysis service.

SELECTIVE HYDROGENATION OF ACETYLENE AND DIENES IN THE PRESENCE OF ETHYLENE AND EXCESS HYDROGEN

The process for the selective hydrogenation of acetylene and dienes (i.e., butadiene) in the presence of ethylene is one of the amazing examples of selective catalysis. It is possible to have a gas stream, for example, of 0.5-1.0% acetylene, 50-70% ethylene, and the balance hydrogen and obtain an effluent containing less than 5 ppm acetylene and zero hydrogenation of the ethylene despite high partial pressure of ethylene and the excess of hydrogen.

The preferred catalyst for the operation is the same as that described as an excellent catalyst for steam hydrocarbon reforming in Chapter 15.

The nickel aluminate–nickel oxide catalyst can be used either as new catalyst or as used catalyst after service in the reforming operation. If the catalyst has been used in the reforming operation it is not necessary to again reduce it, but if it is unused and in the oxide form, it must be reduced using a dilute hydrogen stream (approximately 1% H_2 in N_2, CO_2, or other "inert" gas). The temperature of reduction is 250°C. The H_2 content is increased in stages, after reduction is complete at each H_2concentration. After the nickel oxide is reduced (250°C, 90-100% H_2), it is then sulfided carefully using measured quantities of gaseous H_2S, COS or CS_2 so that the nickel is sulfided to the extent of 20-30% of the total nickel present.

The catalyst is now ready to use at 250–260°C at a gas flow rate of 5000–10,000 space velocity.

Note: The catalyst gradually loses its sulfiding and hence its selectivity, so must be resulfided. Occasionally the feed stream has just sufficient sulfur content to maintain the proper sulfiding level of the catalyst so that selectivity is maintained.

ALTERNATIVE TYPE OF SELECTIVE HYDROGENATION

A catalyst comprising 1–3% CoO, 1–3% Cr_2O_3, and 0.5–2% Fe_2O_3 on alpha-alumina can also be used, but it is more prone to overhydrogenation, so the hydrogen content of the crude feed stream must be carefully controlled. This same type of operation and catalyst can be adapted to service in the selective hydrogenation of acetylene and dienes in pyrolysis gasoline.

19
Dehydrogenation Catalysts

Dehydrogenation catalysts fall into two general categories: simple dehydrogenation catalysts, which act primarily by a high-temperature thermal effect, and oxidative dehydrogenation catalysts, which are the more sophisticated and complicated of the two. In this first section we describe the manufacture of catalysts for thermal dehydrogenation. One of the older processes is the dehydrogenation of butane to butene, which is conducted over a chrome alumina catalyst, the manufacture of which is described below.

Step 1. Alumina hydrate, preferably with about 0.5% Na content and in the form of finely divided powder, is selected and put into a kneader (Fig. 27) or pug mill.
Step 2. An aqueous chromic acid solution (via chromic acid anhydride) in sufficient concentration to derive between 18 and 20% chromium oxide (Cr_2O_3) on the dehydrated

alumina hydrate is now added to the alumina hydrate. The solids and solution should be so balanced as to produce an extrudable paste.

Step 3. The paste can be extruded in equipment such as that in Fig. 28, or it can be dried and then granulated in equipment such as depicted in Figs. 17 or 18 and 29.

Safety Note: Chromium trioxide is a toxic material, and inhalation of the fumes should be strictly avoided. Protective clothing must be used, and ventilation must be well designed.

Step 4. The dried extrudates or pellets are now calcined in equipment such as that shown in Figs. 22, 23, or 26 in air at a temperature of 500–700°C. This brings about some decomposition and some interaction between chromium oxide and aluminum oxide. At this point the catalyst is ready for use. It must first be reduced, but this is usually carried out in place as part of the start-up procedure.

RECOVERY, REUSE, AND REGENERATION OF USED CHROME ALUMINA DEHYDROGENATION CATALYST

Chrome alumina catalyst is one of those for which reuse is the most feasible of the possible recovery procedures for the deactivated material because of the chromium oxide content and the toxicity of the water-soluble Cr^{6+} component. The water soluble Cr^{6+} may be formed as a result of weathering and oxidation of any Cr^{3+} present in a discarded catalyst. It is well to remember that the spent catalyst is not completely deactivated but may become uneconomical to use even when as much as 80% of the original activity remains. This means that regenerating the "spent" catalyst for reuse or recycling may involve only a relatively small percentage increase in activity if the original longevity can be attained or exceeded. This amount of reactivation may be adequately restored by one of the following comparatively simple procedures.

A. Reoxidation in a flow of high temperature (Ca 900°C) steam (hydrothermal) treatment.

B. Item A followed by impregnation with an aqueous solution of ammonium chromate and aluminum hydroxide in which is suspended a gel of aluminum hydroxide. The impregnated particles are dried and then calcined in air at 400–600°C for 2 hr after reaching the desired temperature. The chromate is decomposed to finely divided chromium oxide, which reacts with the aluminum oxide resulting from the dehydration of the aluminum hydroxide. A second impregnation may be desirable to attain the desired level of activity. The aluminum hydroxide acts as an adhesive; the Cr_2O_3/Al_2O_3 ratio may be changed to attain the proper adherence to the original regenerated particles.

C. The reoxidized product of item A is preferably milled to produce a slurry to which CrO_3 and aluminum nitrate are added to convert the slurry to a paste, which is extruded and then calcined to produce a hard extrudate with completely restored activity. Some exploratory tests may be necessary on each lot of spent material to determine the optimum new ingredients necessary to get adequate hardness and activity.

D. If the spent catalyst has been severely abused and cannot be regenerated by any of the above procedures, it too can be handled through a jobber who collects and sells to smelters.

Note: Hexavalent chromium is toxic, and trivalent or divalent chromium weathers to hexavalent. It cannot be used as "fill."

CALCIUM NICKEL PHOSPHATE TYPE BUTADIENE CATALYST

A second type of dehydrogenation catalyst is the calcium nickel phosphate catalyst $Ca_8Ni(PO_4)_6 + 2\%\ Cr_2O_3$ sometimes referred to as the Dow dehydrogenation catalyst. This catalyst can be made as follows.

Step 1. In a kneader of the type shown in Fig. 27, a quantity of 85% phosphoric acid is placed; next powdered calcium hydroxide is added slowly, then nickel oxide (or carbonate), and finally chromic acid anhydride in such a ratio as to produce the composition $Ca_8Ni(PO_4)_6 + 2\% \ Cr_2O_3$. A small amount of additional water may have to be added to obtain the desired paste like consistency.

Step 2. Kneading is continued to ensure homogeneity, usually about 60 min.

Step 3. The catalyst can be extruded (Fig. 28), or it can be dried and pelleted.

Step 4. If the catalyst is to be pelleted, it is, after drying, granulated to 100% through a 10 mesh screen, mixed with 1–2% graphite, and then compressed into pellets in equipment such as that shown in Figs. 29, 31, and 32.

Step 5. After extrusion and drying or pelleting, the catalyst preferably is calcined at 450°C for 2 hr after reaching that temperature. Calcining can be performed in equipment such as that shown in Figs. 22, 25, or 26.

Recovery of Calcium Nickel Phosphate Type
Butadiene Catalyst

Spent calcium nickel phosphate catalyst has nickel values, and calcium phosphate is an ingredient in most fertilizers. The nickel can be processed by a smelter, and if the first recovery step is an acid extraction of the nickel that leaves a nontoxic calcium phosphate, the phosphate can be used beneficially by a fertilizer manufacturer for superphosphate.

DEHYDROGENATION CATALYST FOR ETHYLBENZENE TO STYRENE

The catalyst used to dehydrogenate ethylbenzene to styrene is a mixture of potassium carbonate and iron oxide stabilized with a small amount of chromia. It is prepared by the following very simple procedure.

Step 1. Pigment grade iron oxide or hydroxide is added to a kneading device such as that shown in Fig. 27. Potassium carbonate is then added to the iron oxide to the extent of 18-20% of the weight of the finished catalyst. Chromia equal to 2% of the final weight of the catalyst is added as ammonium chromate or as iron chrome catalyst. Sufficient water is then added to make a paste that is similar to putty but not quite as viscous and not as free flowing as a heavy grease.

Step 2. The wet paste is removed from the kneader and can then be extruded (Fig. 28) or dried in equipment such as that shown in Fig. 22, then is granulated in equipment such as that shown in Fig. 29 and, finally, mixed with graphite in a ribbon mixer such as that shown in Fig. 31.

Step 3. The catalyst is pilled (Fig. 32) to form cylinders or extruded as the earlier prepared wet cake (step 2 above) to produce cylinders, rings, or extrudates of the size demanded for the equipment in question.

Step 4. If desired, the catalyst can be heat treated in air at approximately 500°C for 2 hr for the removal of at least part of the graphite. After this treatment the catalyst is ready for use.

Recovery, Reuse, and Regeneration of Used Styrene Catalyst

The used catalyst may become unusable for several reasons.

1. Carbonaceous material may collect on the catalyst to the extent that it cannot be removed by in situ oxidation. In this case, the carbonized catalyst is removed from the reactor and is segregated into two portions: the most heavily carbonized and that which is relatively clean and retains some activity. New catalyst is put into the reactor with the lightly contaminated catalyst, which acts as a "light-off" catalyst in the reactor, as a "guard" or scavenger for the new catalyst. The most heavily contaminated and deactivated catalyst is carefully reoxidized to remove the car-

bonaceous material at a higher temperature than is possible
in the reactor.

2. The catalyst may not have regenerated to a stable and active condition. In this case it may be necessary to reimpregnate it with a solution of chromic acid and iron nitrate to
derive a fresh surface of iron and chromium oxides. If there
has been some migration or sublimination of the alkali, it
may be necessary to reimpregnate the reoxidized catalyst
with a solution of potassium carbonate. This would best be
done using a pill-coating pan as illustrated in Fig. 37.

3. The catalyst may have been badly abused by a misadventure. In that case it may be necessary to send the reoxidized material to a jobber for the recovery of the chromium,
iron, and potassium carbonate. Because the recovery values
are all low, it may even be necessary to pay the jobber to
take it.

DEHYDROGENATION CATALYST FOR HYDRO-CARBONS AND OLEFINS

The catalyst described in this section is one of a larger family,
but the specific procedure that follows is one that is perhaps
the preferred composition when employing high steam-to-
hydrocarbon ratios (ca. 15:1). The support is unique and noteworthy.

Step 1. Dissolve 8.0 parts by weight of zinc nitrate hexahydrate and 21.0 parts by weight of aluminum nitrate nonahydrate in 40 parts by weight of distilled or deionized
water in a tank such as that shown in Fig. 9.

Step 2. Add the solution prepared in step 1 simultaneously
to another similar but larger agitated tank, with sufficient
28% aqueous ammonia to maintain a constant pH of 7.5
± 0.1 and a temperature of 50°C.

Step 3. Agitate for 2 hr after the completion of precipitation;
then allow the slurry to age without agitation for about
70 hr.

Step 4. Wash by decantation two to four times to remove most of the soluble NH_4NO_3.

Step 5. Spray dry the slurry using equipment such as that shown in Fig. 21 at 200–250°C; next dry at 325°C for 12 hr using equipment such as that shown in Fig. 17.

Step 6. Compress the powder into 6 × 6 mm cylinders using equipment such as that shown in Fig. 32. The zinc aluminum oxide (zinc aluminate) can also be extruded.

Step 7. Crush the pellets in equipment such as that shown in Fig. 29, and then screen it to 8–16 mesh (or other preferable size) in equipment such as that shown in Fig. 36.

Step 8. Calcine the granules at 600–625°C for 3 hr after reaching this temperature in equipment such as shown in Fig. 26.

Step 9. Put the screened material in a coating device such as that shown in Fig. 37 and impregnate the granules with just sufficient volume to completely wet them without excess liquid and with sufficient chloroplatinic acid and stannous chloride to provide 0.37–0.42% platinum and 0.95–1.05% tin.

Step 10. Heat the catalyst preparation in an oxidizing atmosphere to dry it at 200°C for 20 hr using facilities such as shown in Fig. 26. The catalyst is now ready for use.

Other Zn/Al ratios can be used, as can other contents of Pt and Sn. Titanium can also be substituted for the tin. Other variations are also possible and may be preferred for certain hydrocarbons such as alkylaryls.

Recovery, Reuse, and Regeneration of Used
Hydrocarbon Dehydrogenation Catalyst

Note: At high temperatures, 900°C or above, Pt will migrate into the gamma-alumina lattice, where it may be trapped as the alumina condenses to the alpha phase. High temperatures must be avoided, particularly in an environment containing moisture.

Spent or used $ZnO-Al_3O_3 \cdot SnO_2-Pt$ catalyst becomes deactivated for several reasons. For example, the catalyst may be blinded with carbonaceous deposit. The rate of carbon deposition can be minimized by maintaining 1-2% water vapor in the feed along with the hydrocarbon. (**Warning**: in the presence of the water vapor, some hydrocarbon is converted to carbon monoxide and hydrogen in the effluent.) Once the carbonaceous deposit is formed, to regenerate the catalyst, one can pass an inert gas plus 1-2% oxygen over it at 400-600°C to slowly remove the deposit. The catalyst may be regenerated to original values, but carbon deposits are sometimes located within the particle (intergranular deposit) so that when the carbon is removed the catalyst disintegrates to a grainy powder; this, of course, causes unacceptable pressure drop through the catalyst bed. The presence of the precious metal makes it possible to oxidize the carbon with either steam or oxygen at a lower temperature than in the absence of the Pt, for example. Another method for the avoidance of carbon deposition is to incorporate 0.1-1.0% rhenium into the composition.

If the foregoing schemes are ineffective for any reason, the catalyst may be removed from the unit and converter and decarbonized in a separate operation in specific equipment. This may make for a satisfactory regeneration without further treatment.

If further treatment is necessary, it may be possible to accomplish this by spraying or impregnating the catalyst with replacement Pt or $SnCl_2$ or both.

If the spent catalyst fails to respond to any of these treatments, the detoxified spent material can be disposed of through a jobber for metals recovery by an ore smelter.

20
Oxidation Catalysts

OXIDATION OF ETHYLENE TO ETHYLENE OXIDE

The vapor-phase oxidation of ethylene to ethylene oxide is conducted over a supported silver catalyst in a tubular reactor at approximately 230–260°C. The catalyst comprises a low-surface-area, relatively high porosity alpha-alumina or silica alumina support to which is applied silver as an organosilver compound that, on decomposition, will provide 8–15% silver in and on the support. There are many variations of this, such as the addition of an alkaline earth modifier or promoter or the addition of an alkali such as cesium as a promoter or stabilizer and the use of supports of different compositions and physical makeup. It would be impossible to give all of the ramifications, so we will give just the basic process, which can be varied in many different ways, as a search of the patent literature will quickly reveal.

Step 1. A low-surface-area alpha-alumina or silica alumina support having a surface area of less than 0.5 m^2/g and a porosity of 20–24% and in the form of ⅛-in. cylinders is placed in a pill-coating device such as that shown in Fig. 37.

Step 2. A solution is made by reacting silver oxalate with ethylenediamine and ethanolamine as 6% EDA and 4% EA aqueous solution in a ratio of one Ag atom per three EDA and EA molecules to make a solution of the silver. This is made in sufficient quantity for the amount of silver deposited on the support in step 3 to be approximately 13% of the weight of the support.

Step 3. The EDA-EA-silver solution is added to the support and is carefully evaporated to dryness on the support material. (Care must be exercised to avoid the formation of silver azide and its explosive decomposition.)

Step 4. The coated support material is now calcined in a controlled atmosphere using nitrogen, for example, implying the need for a calcining device such as that shown in Figs. 22, 23, or 26. After the calcining operation, the catalyst is ready for use in the oxidation of ethylene to ethylene oxide.

In the above instructions, the use of silver oxalate and EDA–EA solution was described. Instead of the silver oxalate, silver oxide or silver carbonate could be used, and there are still other possibilities. Instead of the aqueous solution of EDA and EA, other amines such as alkynolamines, dimethylamine, and many others would be suitable. However, ammonia itself should be used guardedly because of the possibility of producing an explosive silver azide. The type of support can be alumina as stipulated or a silica alumina such as sillimanite or mullite, both of which are offered by the ceramics industry. The use of beryllia and zirconia as stabilizers has been mentioned in patents. Additionally, barium can be added to the silver slurry in the form of barium oxide, hydroxide, or carbonate, or even a barium halide such as chloride. Other alkaline earths and alkalies can also be used, the quantity (usually less than 1%) and type being at the discretion of the user of the catalyst.

While doing some development work on this EO synthesis catalyst we observed that during the procedure described in steps 1–4 above, there appeared to be a brief synthesis period (about 1 hr) when instead of attaining the usual 65–75% yield, the yield was 85–89%. This was attributed to a rapid reorientation of the silver from its original orientation in the reaction environment to a different one achieved during its initial exposure to reaction conditions. This was somewhat confirmed by X-ray studies, which showed the new catalyst to be amorphous whereas the used catalyst was clearly crystalline (isometric). Attempts were made to avoid or direct crystallization by refractory oxides such as the alkaline earths, silica, titania, or zinc oxide (some of these go through a phase transformation that may be advantageous as temperatures increase), but no method was found to stabilize the orientation in the most favorable state. The lanthanides and the actinides were not studied, and these might prove to be promising. The stakes are high, and a few percent increase in yield could be very valuable.

Another tantalizing finding was reported years ago at the International Congress on Catalysis at Palm Beach, Florida. The statement was made following a paper on the subject to the effect that substituting gold for silver increased the yield to approximately 100% but that the production rate was uneconomically low and did not respond to promoters or activators. One is challenged by the obvious question: Is there a promoter that would make this observation a commercial reality?

Recovery or Reuse of Ethylene Oxide Catalyst

Ethylene oxide catalyst has a very long life, so reuse is not considered advisable. The catalyst is ordinarily returned to the precious metal source, where the silver is dissolved with

with nitric acid and is purified to attain specifications of the catalyst manufacturer: 99.99% silver and less than 10 ppm iron.

OXIDATION OF AMMONIA TO NITRIC ACID

This is a catalyst that is generally used as a wire screen gauze and is manufactured by specialists in this area such as the precious metals companies very likely well known to the reader. Inasmuch as there would be no instructions as to how to prepare the catalyst, instead of instructions, a set of specifications will be given that can be adapted to the industrial application one has in mind or the research objectives. The screen is generally fabricated from 3-mil-diameter wire composed of 90% platinum, with the balance being rhodium and palladium. This is woven into a screen, which may be 80 mesh or coarser. On occasion the screen may be as coarse as 40 mesh and the wire may, under some circumstances, be increased to 6 mils in diameter. The heavier gauzes may be used in the bottom part of the bed to support the lighter 80 mesh gauzes, but the majority of the reaction appears to take place in the upstream portion of the bed, where a catalyst with a relatively open structure is preferred.

The impurity level in the gauze should not be more than a total of 150 ppm, of which less than 50 ppm may be iron. Gold and lead should be avoided.

Occasionally, a problem is encountered in that the gauzes are difficult to light off. This can be dealt with to a certain degree by first coating the precious metal gauze with an adhesive coating of finely divided platinum such as that which could be introduced by coating with a solution of platinic chloride and decomposing it using an oxygen-hydrogen torch.

Other catalysts have been explored for the oxidation of NH_3 to nitrogen oxide for nitric acid synthesis, since costs have been as high as $1000 per ounce for platinum and $8000 per ounce for rhodium. Although this work is noteworthy, it

has not produced a catalyst that gives economically competitive yields. The "other catalysts" have usually been oxides of
base metals such as cobalt and iron and mixtures stabilized
with SiO_2, Al_2O_3, ZrO_2, and MgO. Promoters explored have
been small amounts of the lanthanide, actinide, and alkaline
earths.

Recovery of Values from Used Ammonia Oxidation Catalyst (Nitric Acid Catalyst)

A pack of precious metals gauzes comprising 25–75 sheets is
taken out of service after 3–6 months. It is usually removed
from service because the gauzes have welded together or because the normal growth of dendritic crystals from the wires
constituting the gauze has become so dense and intertwined
that it blocks off gas flow and causes intolerable pressure drop.
Failure can also result from "moth holes" in the catalyst pack
caused by particles of iron rust. When these particles reach
the gauzes, they cause "hot spots" that result in thermal fus-
ing and create a hole through the gauze, with consequent bypassing of unreacted ammonia. This produces a dangerous
situation because of the formation of ammonium nitrate, which
then passes into the nitric acic product. It becomes obvious
that the air must be carefully filtered before passing to the
reactor and the ammonia must be free of iron carbonyl.

Gauze packs that become unstable are returned intact to
the primary metal processor for recovery and reprocessing to
wire and gauze.

During use the temperature of the gauze is high enough
to cause some sublimation of its constituent metals. This metal
vapor is usually recovered on or in a scavenger downstream
of the reactor; the scavenger can be gamma-alumina, gold
screen, or mat. The precious metal vapor is absorbed into the
lattice of the alumina or dissolved in the gold wires. After a
scavenger diminishes in effectiveness, it is processed, by a
smelter, for example, for metals recovery.

OXIDATION OF METHANOL TO FORMALDEHYDE

There are two general processes for the conversion of methanol to formaldehyde. One is identified as the fuel-rich systen that uses silver gauze catalyst (or granular silver) in the form of wire screen mesh or gauze. This catalyst also is manufactured by companies that specialize in precious metals. The second process for the oxidation of methanol to formaldehyde is the fuel-lean process, which uses a molybdenum oxide catalyst. Both processes will be described.

Silver Gauze Fuel-Rich Formaldehyde Process

The silver used in the silver gauze process is what is called five-nines silver, meaning that it is 99.999% silver. It is possible that three-nines silver can be used satisfactorily; this is, of course, left to the judgment of the plant operator or research-er. The gauzes are woven from 0.0006-in. diameter wire to form 18 mesh screen. Heavier wire and screen of different mesh size can, of course, be used, but 18 mesh is the size generally used in commerce.

The catalyst is easily poisoned by iron contamination, which occurs via carryover from the reactant air or the up-stream equipment to the gauzes. This is also true of the nitric acid process previously described; iron carryover via dust is avoided by filtering the air that is used in the process.

It should be pointed out again that it is occasionally diffi-cult to light off the gauzes when they are new, and as a con-sequence, special techniques are used. One method is to pre-treat the gauze by nitric acid etching or to have auxiliary equip-ment that causes hot gases to pass over the gauze before the flow of the air-methanol mixture is initiated. This gives suffi-cient heat and probably etches the gauzes so that they will then initiate the reaction and operate satisfactorily.

As replacement for the gauze, it is possible to use "silver crystals" from electrolysis of silver solutions. The crystalline silver is screened to specific particle size and is usually easy to light off.

The above-described gauzes, both the platinum-palladium-rhodium and the silver, are useful for other operations. The platinum gauze, for example, can be used for the reaction in which ammonia, methane, and air are converted to HCN, and the silver gauze is also used for the vapor-phase oxidation of cyclohexanol to cyclohexanone.

Recovery or Reuse of Spent Silver Gauze

Usually there is no attempt to reuse the gauzes as a pack or as part of a pack, the reason being that the interstices of the wires are blocked with dendritic silver growth or the gauzes have welded together causing intolerable pressure drop. If the gauze has been used for only a short period of time it may be possible to reuse it, but even a short-term use causes the pack to weld together so that when it cools it shrinks away from the reactor, making for bypassing that cannot be compensated for or avoided.

The used pack is almost always returned to a silver smelter for reprocessing.

METHANOL OXIDATION
CATALYST—IRON MOLYBDATE

Although the iron oxide molybdenum oxide catalyst is frequently referred to as iron molybdate, it is really not stoichiometrically iron molybdate. The ratio of iron to molybdenum may be as high as 1 : 5. This catalyst is one of the most difficult to prepare, and deviations from the scheme of manufacture can have a very serious adverse effect on the performance of the catalyst from the standpoint of either selectivity or longevity. It is one of the best examples of selective catalysts that one can point out. Yield can be as high as 95% and conversion as high as 98–99%. The catalyst is also long-lived, lasting for more than a year, and can be reused after that period of time for even more extended service. A preferred procedure for manufacture is as follows.

Step 1. A solution of 1900 parts of water and 80 parts of ammonium heptamolybdate at 30°C is prepared in a corrosion resistant vessel (Fig. 9). In another solution vessel, 60 parts of ferric chloride nonahydrate is dissolved in 600 parts of distilled water, and 60 parts of 38% hydrochloric acid is placed in a third corrosion-resistant vessel.

Step 2. With the agitator operating in the vessel containing the molybdate at a temperature of 30°C, the hydrochloric acid is run into the ammonium molybdate solution rapidly so that not more than 60 sec is required for the addition. After the addition of the hydrochloric acid, the ferric chloride is immediately added at a rate that permits it to be completely added to the acidified ammonium molybdate solution in a period of not over 3 min.

Step 3. A bright yellow precipitate immediately forms. This precipitate and slurry are allowed to agitate at 30°C for an additional 60 min after the precipitation has been completed. The slurry is next filtered in a ceramic or corrosion-resistant filter such as that depicted in Fig. 12.

Step 4. The filter cake is washed by the addition of water to remove the chloride, and simultaneously it is compressed by periodically heavily pressing down on the cake with a wooden mallet. During the course of the washing, the thickness of the cake should be decreased by about 50%. After the washing and filtering operation, the catalyst is dried at 150°C (Fig. 27) and may then be calcined at 500°C (Fig. 26) for a period of 2 hr after reaching that temperature. The latter heat treatment is beneficial to the selectivity of the catalyst.

Step 5. The calcined catalyst is now crushed by passing 100% through a 4 mesh screen using equipment such as that shown in Fig. 29 and is then separated into fractions of 4–8 mesh or 6–10 mesh, as preferred, using a screener of the type shown in Fig. 36. It should be mentioned at this time that pilling or extrusion of this catalyst has not been as successful as leaving the catalyst in granular form.

The catalyst can be modified by replacing some of the iron with bismuth or supplementing the molybdenum with tungstic oxide or vanadium pentoxide. Furthermore, the support can also be supplemented with, for example, a nonreactive form of kieselguhr such as J-M's Hyflosupercel or other similar materials by other manufacturers. Colloidal silica also has a beneficial effect. The use of some of these materials may also make the granules harder, with the result that their tendency to disintegrate in the reactor is minimized. The quantity of colloidal silica can be $5^+\%$ and the diatomaceous earth $10^+\%$. After the granulation and screening, the catalyst is ready for use.

It will be noticed that there is similarity between this catalyst and the bismuth phosphomolybdate catalyst used for ammoxidation and oxidative dehydrogenation, which is described in Chapter 21.

If desired, the wet washed filter cake of step 4 can be extruded using equipment such as that shown in Fig. 28.

Reuse and Recovery of Iron Molybdate

The iron molybdate catalyst produced by the foregoing precipitation procedure has an iron-to-molybdenum ratio of 1 : 5, whereas in the solution from which the catalyst is precipitated the ratio is 1 : 1. The catalyst is identified as an iron molybdate, but it would be more correctly described as a five-member molybdenum oxide cage sequestering a single basic iron chloride molecule.

This structure requires that the ingredients be recovered by dissolving the molybdenum in an aqueous ammonia solution producing water-soluble ammonium molybdate solution and ferric hydroxide, which is removed by filtration. This will be discussed more completely subsequently.

The catalyst is long-lived but eventually becomes inoperable for two reasons. First is that pressure drop in the unit becomes intolerable because of catalyst disintegration and flow blockage by the fines. A second cause of blockage is the sublimation of the molybdenum oxide that crystallizes in the cooler parts of the converter tubes or support wires. The crystallized molybdic oxide apparently derives largely from the exterior of the catalyst particles because when the exterior is removed by abrasion during screening, the interior has the correct iron-to-molybdenum ratio and serves as well as new catalyst in the converter. The purging of low-activity catalyst is apparently adequately achieved by the normal screening operation, which removes the exterior molybdenum-deficient skin.

This granular used screened catalyst is recharged exclusively to the downstream 80-90% of the reactor. The remainder is "topped off" with unused fresh catalyst. Two desirable results are obtained: The fresh catalyst provides a desirable low light-off temperature, and whereas the used catalyst, after its year of operation, had been moderated so that the yields were high and by-products were low, yield reaches 98.5-99% formaldehyde, with conversion being also 98.5-99%.

The fines that are screened from the used catalyst are reused by first calcining and reoxidizing them in air at 500°C for 3 hr after reaching 500°C and then milling them with water and adding them to a fresh batch just before step 2 above.

If the catalyst has been overheated and sintered or fused, it must be reprocessed by dissolving in ammonium hydroxide solution. If it is uneconomical for the catalyst user to do this, the abused catalyst is disposed of through a jobber.

CATALYST FOR THE OXIDATION OF SULFUR DIOXIDE TO SULFUR TRIOXIDE

This catalyst also fits into the so-called oxide or mixed oxide type of oxidation catalysts. It will be noted that there is a strong similarity in composition between the catalyst used for the oxidation of sulfur dioxide to sulfur trioxide and those catalysts that are used for the synthesis of maleic anhydride and phthalic anhydride.

This catalyst family is unique in many respects, one of the most unusual being that it is often used in the molten form. This is reasonable because vanadium pentoxide and potassium pyrosulfate form a eutectic mixture that melts at a temperature approximately the same as the reaction temperature in the sulfuric acid converter (550–650°C). The catalyst is entrapped in the matrix of silica and kieselguhr, and the vanadium pentoxide and potassium pyrosulfate are molten islands in this silica matrix. This catalyst is extremely long-lived and may perform for as long as 10 years before it gradually disintegrates to powder. The catalyst is restored to usefulness by annually removing it from the converter, screening it, and replacing in the converter the portion that is retained on a 6 mesh screen, for example. Its preparation follows.

Step 1. A quantity of kieselguhr or diatomaceous earth is added to a kneader such as the one depicted in Fig. 27.

Step 2. To the kneader is also added an amount of potassium sulfate equivalent to 60% of the weight of the kieselguhr. Next are added portions of cobalt sulfate, nickel sulfate, and iron sulfate, each equivalent to 2% of the weight of the kieselguhr. Next is added sufficient ammonium metavanadate to be equal to 20–30% of the initial weight of the kieselguhr.

Step 3. Sufficient distilled water is added to produce a paste. The kneader blades are started as soon as sufficient water has been added to lubricate the powders and permit the formation of a pastelike mass. Kneading is continued until the paste becomes uniform and has a texture not quite as viscous as putty but slightly more viscous than a heavy grease. When this point has been reached, generally after a period of 45 min to 1 hr, the mixing should be continued for an additional 10 min as insurance of thorough mixing and uniformity. The kneader is then discharged.

Step 4. The kneaded catalyst is dried and heated to a temperature of 300°C in equipment similar to that shown in Figs. 17, 22, and 23.

Step 5. After drying and slight calcining, the catalyst is broken into lumps and passed 100% through a 10 mesh screen, using a granulator of the type shown in Fig. 29. The catalyst is now mixed with 1% graphite and is pilled in equipment of the type shown in Fig. 32. The ribbon mixer for the graphite mixing is shown in Fig. 31. The catalyst is now ready for use. This catalyst can also be extruded in equipment such as that shown in Fig. 28, or it can be formed into spheres using equipment of the type shown in Fig. 33. The primary requirement of the catalyst is that it have strong durability (resistance to abrasion) under long periods of exposure.

A large number of modifications in this catalyst can be made, again depending upon the needs of the laboratory researcher or the plant operator. The catalyst can be hardened by replacing some of the kieselguhr with a colloidal silica such as DuPont's Ludox SM or one from another supplier. It is possible that titania could be added in small quantities, possibly as Tyzor (DuPont), but both of these, particularly the colloidal silica, will tend to result in much harder extrudates or spheres produced by the spherudizer.

Some patents claim that alkali metals, other than potash or together with potash, are beneficial. Rubidium and cesium have both been mentioned. There is some evidence also that any means of increasing the surface area of the material would be highly beneficial, and it is also possible that any method of increasing porosity would be highly beneficial. These are only mentioned on a speculative basis; it is possible that none will improve the catalyst, but the patent literature claims that substantial improvements could be made. It is true that if a catalyst could be devised that had a lower operating temperature, it would be very valuable and useful, because it could be used to achieve higher conversion at a lower temperature with the consequent minimization of loss of SO_2 in the exit gas (and into the atmosphere).

Reuse of Used SO_2-to-SO_3 Oxidation Catalyst

The SO_2-to-SO_3 oxidation catalyst has an unlimited life measured roughly as 10 years because the catalytic reactor is shut down each year for maintenance. During the shutdown the charge of catalyst is removed from the multiple-tray reactor, screened, and recharged. There generally is a 10% volume loss to fines, which is replaced with new catalyst, which in turn is used exclusively on the downstream (bottom) tray or stage. The new catalyst is used in the final stage so that this stage can be operated at the lowest temperature, favoring maximum conversion and minimum leakage of SO_2.

In the past, fines were returned to jobbers for vanadium recovery, but more recently there appears to be economic advantage in regeneration. Several regeneration schemes have been proposed.

Scheme 1. Reoxidation in air at a temperature of 350°C to as high as 600°C. Reform into granules, pills, or, after milling with water, extrudates.

Scheme 2. Scheme 1 or 4 (below) plus addition of the slurry to a new catalyst preparation following precipitation step 3 using the slurry to replace the water otherwise called for in the kneading operation.

Scheme 3. Scheme 2 with a hydrothermal treatment of the slurry of reoxided used catalysts and the fresh catalyst. Hydrothermal treatment may be at atmospheric pressure at 75-100°C or at elevated temperature and pressure. Corrosion may be severe, so this factor must be taken into account in the design. If a boost in activity is required, it may be desirable to add to the slurry 1-2% additional NH_4VO_3 as promoter; 1-2% cesium carbonate also should help activity but should be added during the subsequent kneading. If additional hardness is required, 2-10% (as solids content) of Tyzor, colloidal silica, or colloidal ceria can be added during kneading.

Scheme 4. This is a procedure devised with the idea in mind that the used catalyst is in a reduced condition and reoxidation in air may entail an exothermal reoxidation that could do permanent harm to the catalyst, making the regeneration futile. The used catalyst is reoxidized while still in the converter (or an auxiliary converter) using a gas comprising N_2 or CO_2 containing only 0.1% O_2. The O_2 is exhausted from the gas by the reoxidation of the catalyst. The temperature is maintained below 500°C and must be carefully maintained. As the heat evolves (temperature rise is no longer evident) the O_2 content of the gas (usually recirculated) is increased, to 0.5% for example, and the temperature rise is monitored. The oxygen content of the gas is increased in small increments until pure air is being introduced. The temperature of oxidation should be held as low as possible and should never exceed 600°C.

As a last resort the unregenerated catalyst can be transferred to a jobber for metals values recovery. This is usually a low-value outlet, so regeneration schemes are economically most attractive.

OXIDATION CATALYSTS FOR THE SYNTHESIS OF MALEIC ANHYDRIDE AND PHTHALIC ANHYDRIDE FROM BUTENE AND NAPHTHALENE

These catalysts are basically the same as the catalysts used for the oxidation of SO_2 to SO_3. However, in some cases the process calls for a fluidized bed, in which case the catalyst is processed in slurry form through a spray dryer such as that described in Fig. 20. When the catalyst is processed through a spray dryer, it is highly desirable that hardening materials such as Ludox or another silica colloid be used to improve the strength to increase abrasion resistance. Other than the fact that the catalyst is, in some cases, used in microspheroidal form, the composition is essentially the same as that used for the sulfuric acid process.

Maleic Anhydride from Butane

This process employs a vanadium oxide–phosphorus pentoxide catalyst sometimes promoted with MoO_3, WO_3, Sb_2O_3, TiO_2, TaO_2, or NbO_2 supported in SiO_2, TiO_2, ZrO_2; alkalies and alkaline earths or mixtures of these also are frequently added (see References). This catalyst can sometimes be used advantageously as a replacement for the SO_2-to-SO_3 type of catalyst previously described for butene oxidation to maleic anhydride. The preparation is as follows:

Step 1. One hundred parts by weight of V_2O_5 and 36 parts of $ZrOCl_2 \cdot 4 H_2O$ are added to 1200 parts of 37% HCl in a tank such as the one in Fig. 9.

Step 2. This solution is heated and refluxed for 2 hr; then 153 parts by weight of 85% H_3PO_4 is added. Refluxing is continued for another 3 hr.

Step 3. The resultant gel is removed from the vessel, dried at 150°C, and then granulated and screened to use as granules or to mix with 1% graphite and pill. The catalyst also can be supported on Al_2O_3, $SiO_2 \cdot Al_2O_3$, etc. The catalyst can also be spray dried or extruded. **Warning:** It is extremely corrosive.

Variations of the catalyst can be made by varying the ratios of V, Zr, and Ti and also by substituting Cr, Fe Hf, La, or a rare earth for the Zr.

Reuse and Recovery of Used Maleic Anhydride and Phthalic Anhydride Catalyst

The procedure for recovery or regeneration of used MA and PA catalyst is very similar to that described for the SO_2-to-SO_3 catalyst, with one exception. The MA and PA catalysts may be contaminated with carbonaceous residue from the selective oxidation, and the vanadates may also be at least partially reduced to vanadite. Each of these conditions requires a careful reoxidation with a gas comprising initially 0.1–0.5% O_2 in an inert gas such as N_2 or CO_2. As the oxidation is completed at any given O_2 level, the O_2 level is increased to 21% (pure air).

21

Ammoxidation: Synthesis of Acrylonitrile from Ammonia and Propylene

The catalyst for the synthesis of acrylonitrile from ammonia and propylene is prepared in a way similar to that used for sulfuric acid synthesis catalyst described in Chapter 15. It is used in a fluidized bed and as a consequence is processed by spray drying. The process is as follows:

Step 1. A quantity of ammonium molybdate is added to a kneader of the type shown in Fig. 27. (The equipment must be corrosion-resistant.) Sufficient bismuth nitrate is also added to the kneader to give a ratio of 2 atoms of bismuth for each 3 atoms of molybdenum. Also added to the kneader is ferric nitrate equal to 1/10 of the atomic equivalent of the molybdenum and potassium carbonate equal to 5/100 of the atomic equivalent of the molybdenum. Finally, there is added Ludox SM with a silica content equal to the total oxides content of the ingredients already present. The kneader is started, and sufficient ammonium bicarbonate is added to be equal to the stoichiometric equivalent of the cations of nitrate present in the mix. This action tends to thicken the material and cause a gelling action, which in turn benefits the uniformity of the catalyst.

Step 2. The mixing is continued until a uniform appearance has been attained. Generally, this is approximately 15 min.

Step 3. At this point, the slurry is spray dried (Fig. 20) to produce microspheres that are essentially in the range of 60–200 mesh. Other ranges can, of course, be selected and used.

Step 4. The microspheres are now calcined in air (Figs. 22, 25, and 26) at a temperature of 500°C for 2 hr after the catalyst reaches that temperature. [Nitrogen oxides (NO_x) are evolved, so abatement facilities must be provided.] After the calcining operation, which may also be at a lower or higher temperature (up to 700°C), depending upon the preference of the researcher or the operator of the process plant, the catalyst is ready for use.

Modifications can be made in many respects. For example, nickel can be added along with the iron as a cocomponent of the catalyst. Other types of Ludox can also be used, including even a smaller size than the 80-A Ludox SM. A higher or lower proportion of the Ludox can be used, and titania can beneficially be substituted for some or all of the silica. Zirconia from zirconium nitrate or colloidal zirconia also can be employed beneficially.

This catalyst, although stipulated for use in ammoxidations, can also be used for oxidative dehydrogenations such as the dehydrogenation of butene to butadiene and also for the oxidation of propylene to acrolein and other alpha-positioned olefins to the corresponding aldehydes.

Reuse, Regeneration, and Recovery of
Ammoxidation Catalyst

There are probably few if any catalysts that have received more research both on modified compositions of the new catalyst and on the bonding species in both the new and used catalyst. There are literally hundreds of papers and articles on the subject. Because the catalyst is heated during fabrication at temperatures approaching 700°C, there are, in addition to the species formed during preparation, solid-state reaction and with composition containing BiO_x, MoO_xPO_x, SiO_2, FeO_x, and TiO_2. Since there are also alkali hydroxides or carbonates that are mineralizers and lanthanides, it is immediately evident that the product species can be unusually complex and variable. Adding to those variables is the fact that the catalyst is fabricated in an oxidizing environment but is used in a redox environment. The confusion and contradictions in the literature are understandable.

In addition to the broadly used bismuth, molybdenum, iron, potassium, and silicon oxide catalyst, Nitto has developed and used a catalyst made up of tellurium, antimony, iron, potassium, zirconium, molybdenum, and silicon oxides. There are many patents by Nitto and, of course, variations in the composition using other oxides from the same elemental family, that is, Li and Na for the K. High yields are claimed for this catalyst family usually designated the tellurium family.

Regeneration of the ACN Catalysts

There are many schemes used for the regeneration of the used (not spent) catalyst. It is reported that the catalyst has unlimited life if it is regenerated by one of these methods.

The basic method involves the replacement of the molybdenum that analysis of used catalyst indicates has sublimed from the catalyst particles. The Mo is replaced in the form of a spray, mist, or vapor containing MoO_3 as the ammonium salt, the paramolybdate, or the trioxide. If the replacement is properly performed during use or on a side stream to be recycled, the yield can be restored to standard, or if the sidestream undergoes constant withdrawal and treatment the performance of the catalyst can be maintained at a constant high level.

In the case of the tellurium catalyst, the procedure is the same except that the element to be replaced is the tellurium, although it may be necessary to also replace the P_2O_5. Molybdenum is not reported to be fugitive, but after long-term use, replacement of some of the molybdenum may also be necessary.

Reclamation of the Metal Values from Spent Catalyst

It does not appear to be necessary to reclaim metal values since the catalyst seems to respond so favorably to regeneration procedures that it does not become spent.

22

Oxychlorinations or Oxyhydrochlorinations

The catalysts for oxy- or oxyhydrochlorinations are ordinarily supported copper chloride. Members of this family of catalysts are generally high-surface-area alumina impregnated with copper chloride. These are made following the general method of impregnation via the pill-coating scheme already described.

Step 1. Charge to the corrosion-resistant or rubber-lined pill-coating machine (Fig. 37) a quantity of activated alumina in granular form, and then charge an aqueous solution of copper chloride sufficient to produce a 6–14% quantity of copper chloride on the basis of the quantity of alumina in the charge.

Step 2. Rotate the pill-coating machine pan and the contents to ensure that the catalyst is uniformly coated, adding water if necessary or heating the jacket to drive off excess water.

Step 3. After the "drying" operation at 175–250°C (Fig. 18), the catalyst is ready for use.

This is a rather general way of preparing chloride catalysts, this one happening to be copper chloride on granular or spherical alumina, silica, or microspheres of either silica or alumina. Further, mercury chloride on support(s) can also be prepared this way for use in the manufacture of vinyl chloride from acetylene and HC1. It is apparent that other modifications and other types of catalysts can be made by substituting chromia, titania, or zirconia for the alumina or silica supports. (One must be careful to avoid forming a volatile chloride from the support.) A catalyst could very likely be derived that would be suitable for the synthesis of fluorocarbons from their respective fluorochlorocarbon sources.

Safety Notes

Activated alumina has, as the word "activated" indicates, an unusually strong tendency to absorb gases, vapors, and liquids. As a consequence, the new catalyst must be stored in gastight containers, and the following precautions must be taken during its removal from the containers and charging to the unit.
    Maintenance personnel must use:

1. Respirators for dust – absolutely essential
2. Gloves to protect hands – rubber or canvas as deemed necessary, preferably rubber
3. Clothing to protect the body to the extent necessary, including avoidance of allergic reaction
4. Air pack if conditions are that severe

When the catalyst is ready to be discharged, the following must be considered:

1. Is the catalyst contaminated with carbon, and is it reduced so that it is pyrophoric, oxidizable, combustible? If so, it

must be oxidized in the reactor to avoid serious fires either at site of use or at a recovery plant. Quenching in water is not recommended.

2. Is the spent catalyst contaminated with toxics; skin, eye, or nasal irritants; carcinogens; or odorous materials? Special handling must be practiced with each of these conditions or combinations of them. Each is serious and can have expensive consequences as well as causing trouble with OSHA or EPA.

3. The spent catalyst may vary in the contaminants listed in item 2, above, because of variations in feed purity and operating conditions. Such changes must be monitored carefully and taken into account in the care exercised in handling the spent catalyst.

Reuse, Regeneration, and Recovery of Oxychlorination or Oxyhydrochlorination Catalysts

As this is being written, there are several vocal groups calling for laws forbidding the use of any organic chlorides in commerce. Although it is deemed unlikely that such severe restrictions will be imposed, it does give an indication of the rather general concept that all organic chlorides are suspect of being carcinogenic. In light of the possibility of carcinogenicity and the factors discussed under "Safety Notes," it is obvious that extreme care must be exercised in removing the catalyst from the reactor, placing it in containers, storing it, and disposing of it.

It is likely that the preferred route of disposal is through a professional agency that is familiar with such hazards and is equipped with both the knowledge and the facilities to deal with such materials. A tabulation of companies can be easily obtained for any given area in the telephone directory Yellow Pages under Waste Reduction, Disposal, & Recycling.

23

Vinyl Acetate Synthesis from Ethylene and Acetic Anhydride

This is an interesting and rather unique catalyst in that it comprises a combination of both an electron-deficient element-palladium, iron, or the like – and electron donor(s) such as silver or copper. The method of preparation is as follows.

Step 1. Place in a pill-coating device (Fig. 37) a quantity of silica in the form of high-surface-area pills – 0.25-in. spheres, for example, or cylinders. To this silica, add sufficient 10% sodium hydroxide solution to wet the silica in the spheres uniformly but not excessively. Remove the wet spheres from the pill-coating device and put them into a corrosion-resistant open-top tank (Fig. 9) with a diameter such that the entire product from the pill-coating device can be accommodated in the bottom of the vessel at no more than a 2-in. depth. Now add to the support a solution that just covers the alkali-treated silica support. This solution should contain an amount of silver nitrate equal to 4% of the weight of the silica as metallic silver and palladium nitrate equivalent to 1% elemental palladium. Let the solution stand for 24 hr, during which time the Pd and Ag will be deposited on the outer skin of the support material. It is possible that a shorter time can be used if the observation is made that the palladium and silver are completely precipitated from the solution.

Step 2. Either in the same vessel or in another vessel, wash the coated catalyst with distilled water until the sodium and nitrate have been essentially completely removed.

Step 3. Remove the catalyst to drying trays (Fig. 22, 25, or 26) and dry it at 150°C in air. This drying completes the preparation of the catalyst. It is now ready for use.

The silica support stipulated above is manufactured and sold by Sud Chemi, a branch of Lurgi in Germany. It can be obtained in this country from the affiliate of Sud Chemi, United Catalyst of Louisville, Kentucky. Instead of the silver stipulated, other elements such as iron, copper, and gold can be used, but the effectiveness of the resulting catalyst is somewhat inferior to that of the catalyst that contains silver. Palladium and silver in concentrations other than those stipulated can also be used, but this is dependent upon the researcher's own findings for the particular operation and on how these factors affect both yield and productivity per unit volume of the reactor.

REGENERATION OR METALS RECOVERY FOR THE SPENT VA SYNTHESIS CATALYST

The catalyst ordinarily is long-lived and is deactivated by organic polymer-like deposits, which can be removed by careful reoxidation to remove the carbonaceous material. Because silica sinters at a relatively low temperature, care in reoxidation is emphasized.

Metals recovery is achieved by simple reoxidation and acid extraction.

Safety Note. Extreme caution must be exercised in the handling of the spent catalyst, using safety clothing and breathing facilities.

This particular catalyst and process should cause no toxic materials to form on the catalyst, but this does not mean that some abnormal conditions could not be encountered, producing questionable deposits of unknown properties. The basic rule should be to suspect the worst and be prepared for it. It is well to keep in mind the rhetorical question that was the title of a lead article in a popular magazine recently: "Is Everything Carcinogenic?"

24

Supported Precious Metals Catalysts

Supported precious metals catalysts are all considered together for the reason that they can be divided into individual operations. Such catalysts, for example, are used for hydrogenations, oxidations, oxychlorinations, dehydrogenations, and oxidative dehydrogenations as well as for other processes. They also are useful for the hydrogenation of organic compounds containing halide groups as well as sulfur and phosphorus. These catalysts can be operated at a low enough temperature that the hydrogenation can be carried out without hydrogenolysis and the removal of these important substituent groups or components.

Although precious metals are usually grouped into a class, implying that each precious metal behaves very similarly to any other precious metal, this is not the case. As an example, chlorine can be stripped from palladium at quite low temperatures with hydrogen. This, however, is not true of ruthenium.

Ruthenium probably is the precious metal that deviates far-
thest from the precious metals grouping. Furthermore, ruthenium has very specific properties that make it highly attractive
for certain hydrogenations and even oxidations. As an example, ruthenium behaves much the same as rhodium in the three-
way catalyst for automotive exhaust abatement, with the exception that ruthenium forms a volatile oxide that is both toxic
and volatile. As a result, the catalyst is quickly stripped of
the ruthenium in an oxidizing atmosphere at a typical temperature of automotive exhaust abatement and is consequently
unsatisfactory for that application.

In our description of the preparation, we deal with palladium and platinum, but in essentially all cases other metals
can be similarly processed with suitable, usually minor, changes
in preparation and use conditions.

A word of warning is worthwhile when discussing and
suggesting the use of activated carbon, which is frequently
used as a support for precious metals. Activated carbon is the
most difficult material to specify, primarily because of its natural origin. For example, the ash content of anthracite coal
is extremely variable even from seam to seam in the same mine,
the ash content of coconut char also varies depending upon
the location of the coconut trees and the amount of rainfall
experienced during the year.

REGENERATION AND METALS RECOVERY

The regeneration and recovery of the precious metals are similar, so they are presented at the conclusion of Chapter 30 (pp.
259–261).

25

Palladium on Powdered Carbon

Carbon particles generally are basic, and if they are sufficiently basic, it is unnecessary for them to be pretreated in order to effect precipitation of the precious metals. However, if the precious metals content is above 5%, it is preferable and safest to pretreat the catalyst support with alkali to expedite and make complete the precipitation.

Step 1. A 1% (on the basis of elemental palladium) solution of palladium chloride or palladium nitrate is prepared (Fig. 9) of a sufficient volume to produce the quantity of precious metals that is preferred on the activated carbon, that is, 0.1–5%. A slurry of the activated carbon to be coated is prepared that contains 1–5% activated carbon solids in water and is rapidly agitated. The slurry is adjusted to 50°C.

227

Step 2. The palladium solution is also maintained at 50°C, and the pH is increased with an alkali hydroxide or carbonate (e.g., NaHCO) from the normal 0.1 pH to approximately pH 4.0, which is just to the point of incipient precipitation of the Pd.

Step 3. The palladium solution with pH adjusted is now rapidly agitated, and the slurry of the activated carbon is rapidly added and the pH noted. Because of the inherent basicity of the activated carbon, the pH should have risen to 5.5–7.

Step 4. With the slurry continuing to be rapidly agitated, sufficient 10% sodium bicarbonate solution is added to bring the pH up to 7.0 ± 0.2.

Step 5. The slurry is allowed to agitate for an additional 30 min and is then allowed to settle and is washed by decantation using distilled or demineralized water until the halide and sodium ions have been adequately (essentially completely) removed.

In some cases, ammonium bicarbonate is preferred to sodium bicarbonate in step 2, and in others the hydroxides are preferred. In other cases, palladium nitrate is preferred and the sodium is removed in step 5, but not as completely as when the halides are present. Instead of adding the carbon powder as slurry (steps 1 and 5), it can be added as a dry powder if preferred.

The palladium catalyst must be reduced before being used for hydrogenation, and this is accomplished by any of several different techniques. The simplest involves adding formaldehyde in 10 times stoichiometric quantity to the slurry, which effects the reduction of the metal as the slurry temperature is raised to 100°C.

The catalyst can also be reduced by bubbling hydrogen through the slurry or by first filtering the slurry, removing the catalyst from the liquid, and then placing the catalyst in a reducer and passing hydrogen over it to both dry and reduce it. Reduction in this case can be effected at temperatures from

80 °C to approximately 200 °C. Temperatures above 200 °C should be avoided.

Note: After reduction, the catalyst might be pyrophoric (ignite spontaneously) when dry and must be protected from exposure to air, either by keeping it in an inert atmosphere or by keeping it immersed in a suitable liquid such as the product of hydrogenation or the product to be hydrogenated.

REGENERATION AND METALS RECOVERY

Regeneration of palladium or other precious metals on carbon is very difficult, and no general instructions are possible. If the deactivating agent is a wax, polymer, or high-boiling organic, the first scheme to try would be solvent extraction because the carbon support would also oxidize during an oxidation of the organic contaminant.

Safety Note: The warnings given previously regarding handling of other spent catalysts should be noted and observed. To repeat: Spent catalysts can be contaminated with unsuspected toxic materials.

PALLADIUM ON GRANULAR CARBON

The procedure for preparing a catalyst of Pd on granular carbon is almost identical to that described above for powdered activated carbon, but the agitation must be carefully carried out to avoid breaking the carbon granules, and particularly to avoid abrading the surface-deposited palladium from the granules. It may also be preferable to pretreat the activated carbon with alkali in order to effect better precipitation of the palladium on the surface.

The process description given in a subsequent section for the impregnation of alumina is in sufficiently general terms that it can be adapted to Pd on activated carbons. It should always be borne in mind that both aluminas and carbons, particularly carbons, are variable, and some process adjustments may be necessary from lot to lot of these supports.

PALLADIUM ON SPHEROIDAL OR GRANULAR ACTIVATED ALUMINA

We first reemphasized that aluminas, like activated carbons, are as variable as the weather. Activated carbon is particularly unspecifiable and unpredictable, but alumina is often not recognized as being as variable as it really is. First, there are a number of different species of alumina, and when one defines a catalyst as supported on alumina, it really means nothing because alpha-, gamma-, eta-, etc., aluminas all behave very differently from one another. There is even a difference between spheres of the same alumina species produced by different vendors.

One of the major problems with alumina spheres is that they often decrepitate when dropped into water. "Decrepitate" refers to the breaking away of the exterior portion and the almost complete disintegration of the sphere into particulate matter. (Granular material does not seem as susceptible to this problem.) This decrepitation can be avoided by preliminary moisture treatment of the spheres with steam in the vapor phase at approximately 125°C. Decrepitation does not usually occur after the spheres have been rehydrated to a point of approximately 50% saturation.

To avoid this need for humidification, however, it is best to purchase spheres that do not have this objectionable property. Most alumina supply companies have spherical products that will not decrepitate. It can be seen that when a decrepitation-prone catalyst is used in a given operation and some misadventure causes the catalyst to become wetted with water or even an organic liquid, decrepitation and complete destruction of the catalyst charge may result.

Regeneration or Metal Recovery

This subject is presented at the end of the section Regeneration or Recovery of Automotive Exhaust Catalyst in Chapter 30.

PREPARATION OF PALLADIUM ON ALUMINA SPHERES

Step 1. The total pore volume of alumina spheres is determined by weighing a quantity of the spheres, immersing them in water and allowing them to remain immersed until they are completely saturated, removing the spheres from the water, blotting their surface dry, and finally weighing them again. The difference in weight is the total pore volume of the alumina spheres.

Step 2. The spheres are made alkaline by preparing a 1–5% sodium bicarbonate solution and using an amount of this solution exactly equal to the total pore volume of spheres involved.

Step 3. The spheres to be treated are placed in a pill-coating device of the type shown in Fig. 37, and the volume of sodium bicarbonate solution corresponding to the total pore volume is then poured into the device.

Step 4. After complete coating, the spheres are dried, preferably in the pill-coating device itself (steam-heated or electrically heated coater).

Step 5. A solution of palladium chloride is prepared that (1) contains the proper quantity of palladium to be impregnated onto the surface of the spheres and (2) is equal to the total pore volume of the spheres already determined.

Step 6. The spheres, while being rotated, are coated with the exact volume of palladium chloride (or nitrate) solution stipulated in the previous step. Rotation is continued until the spheres are completely and uniformly coated.

Step 7. Rotation is continued for an additional 1–20 min, and then the spheres are washed to remove the halide and sodium by running distilled or demineralized water into the rotating pill-coating device and allowing it to overflow until such time as the chloride and sodium are at the specified low levels.

Step 8. The spheres are removed from the pill-coating equipment and are reduced in liquid phase at 90°C using formaldehyde as 5% solution in a volume of water equal to 10 times the quantity of hydrogen necessary to reduce the theoretical palladium oxide. The catalyst is now ready to be used in a reducing environment. If, however, the catalyst is to be used in an oxidizing environment, it is unnecessary to reduce it. For low-temperature oxidations, platinum or rhodium are frequently considered to be superior to palladium.

Regeneration or Recovery of Metals
from the Spent Catalyst

See the comments and instructions at the end of the following section.

PALLADIUM ON POWDERED ALUMINA

The supporting of palladium on powdered alumina is much like the process previously described for nickel on kieselguhr or alumina but is given here because there are some slight variations that should be noted.

Step 1. The desired quantity of palladium as palladium chloride or nitrate is dissolved by agitation in a corrosion-resistant piece of equipment such as that shown in Fig. 9. The concentration of the palladium is generally in the range of 1% (calculated as elemental palladium), and the temperature is usually between 40 and 60°C.

Step 2. A 10% solution of ammonium bicarbonate or sodium bicarbonate is now slowly added to the solution of palladium salt in the precipitation tank. The pH is increased to the point of incipient precipitation, which should be approximately pH 4.0.

Step 3. Powdered aluminum oxide or alumina hydrate in the desired species of gamma- or eta-alumina or alpha- or beta-alumina hydrate is rapidly added to the tank containing the palladium salt with adjusted pH.

Step 4. The pH is further adjusted to 7 ± 0.2 by further addition of the precipitant as a 10% solution.

Step 5. Digestion is continued for an additional 60 min at about 60C, and then reduction is effected by the addition of formaldehyde or other reducing agent in a quantity sufficient to be equal to approximately 10 times the stoichiometric amount required for the reduction of the palladium oxide. Hydrazine can be used as reducing agent, but the catalytic characteristics differ from those of the catalyst reduced with formaldehyde.

Step 6. The slurry is washed by decantation or on a filter to remove the sodium and halide to an acceptable level, which usually means just barely detectable in the wash water.

Step 7. The catalyst is filtered and stored in the wet condition unless the dry form is preferred. It is generally dried in a vacuum or controlled atmosphere oven (Fig. 26), but if the catalyst is to be used in an oxide state (oxidation) and was not reduced in step 5, it can be dried in an air or oxidizing atmosphere.

Many variations on the above procedure can be performed such as substituting platinum, rhodium, ruthenium, osmium, or iridium for the palladium stipulated. Each of these metals has somewhat different characteristics and characteristics that are particularly attractive for certain operations. Mixtures of the metals are sometimes preferred to the individual elements. Platinum and palladium are frequently used together, and platinum and rhodium or palladium and rhodium. Iridium and rhodium frequently are strong promoters for either palladium or platinum.

Safety Note: Osmium can form a volatile salt that can coat one's eyes, causing irreversible blindness.

Precipitants can be sodium carbonates or hydroxides or ammonium carbonates or hydroxides. Also, instead of the formaldehyde reductant, one can use hydrazine, which gives a distinctive catalyst quality, sometimes superior but sometimes inferior to catalyst reduced with hydrogen or formaldehyde.

There are several truisms relative to the placement of the precious metals on both the granular and powdered types of support. It is generally true that a greater percentage of metal can be economically used if the catalyst is in the form of powder. This concentration may go as high as 5%, but in the case of a spherical support, for example, generally 0.5-1.0% is entirely adequate and above that the metal is being wasted. It is generally true that the smaller the crystallite size of the precious metal that is developed, the more active the catalyst; the use of the carbonate as precipitant is generally preferred to the use of the hydroxide for this general reason. It is also preferred to hold the temperature as low as possible during precipitation to minimize crystallite agglomeration or growth. There are probably other trade secrets that are employed to obtain marginal effects.

Regeneration or Metals Recovery

This subject is covered fully at the end of Chapter 30 (pp. 259-261).

26
Enantioselective Catalysts

The application of enantioselective catalysts in the pharmaceutical and agrichemical industries has increased rapidly over the last two decades. The trauma associated with the use of racemic thalidomide led regulatory agencies to demand vigorous justification for approval of the use of a racemate in products for biological work. The work of W.S. Knowles et al. showed that the use of a homogeneous organometallic compound with a chiral ligand could give product of sufficient optical purity (>95%) to be economically attractive. The development was based on the use of the homogeneous hydrogenation catalyst of Wilkinson [$RhCl(PO_3)_3$], replacing triphenyl phosphine with optically active phosphorus, to reduce a prochiral olefin. The high optical purity product results in greatly reduced investment in synthesis and purification, which compensates for the expense of the catalyst, which further contributes a catalyst multiplier factor.

There has been continuing effort in seeking enantioselective heterogeneous catalysts that offer the technical advantages of catalyst separation and recycling. However, only a very few heterogeneous enantioselective catalyst systems have been found, and they are usually substrate-specific. The dominance of homogeneous catalysis is tied to the application of organometallic synthetic chemistry that permits systematic optimization of structures to give not only the desired optical selectivity but also the desired process attributes of rate and stability. Systematic studies of the hydrogenation of the α-acylamino acrylic and cinnamic acids elucidating the reaction mechanism and the effect of structure of the chiral phosphine cationic rhodium catalyst on the optical purity of the product provided a firm foundation for continuing development. These studies showed that the more thermodynamically stable complex of the optically active phosphine cationic rhodium catalyst and the prochiral substrate produced the minor, undesired, optical isomer. A large number of reactions using optically active metal catalysts have been reported. Most of the reactions operate on the conversion of olefinic or carbonyl functions and cover hydrogenations, epoxidations, and C-C bond formation.

The prime example of commercial enantioselective catalysis is the Monsanto L-dopa process. The process is the hydrogenation of a substituted α-acetamidocinnameric acid with a catalyst based on rhodium complexed with an optically active phosphine. The process evolved from the initial use of a monodentate chiral phosphine (cyclohexylmethyl phosphine) to bidentate phosphines that give more stable complexes and permit fine tuning of the catalyst activity and process attributes.

The preparation of the enantioselective catalysts is an exercise in organometallic chemistry. The quantities of catalysts required are generally small (10^2–10^4 lb/yr) and involve the use of both an expensive metal and expensive reagents. The problems in scale-up of a demanding synthesis has frequently led to having the producer also manufacture the catalyst. A publication by Chan et al. describes a simplified catalyst preparation. The standard synthesis consisted of two steps. The first step was the reaction of a purchased ruthenium cyclooctadiene dichloride with a purchased phosphine, S-BINAP, in the presence of excess triethylamine to form the ruthenium BINAP dichloride triethylamine complex. In the second step the product is reacted with sodium acetate.

Step 1. $[Ru(COD)Cl_2]_n$ + BINAP⟶$[Ru(BINAP)Cl_2]_2NEt_3$
Step 2. $Ru(BINAP)Cl_2]_2NEt_3$ + xsNaOAc⟶Ru(BINAP)$(OAc)_2$

In an attempt to simplify the synthesis, the reaction was carried out in acetic acid with sodium acetate, thus eliminating the use of triethylamine. The product was obtained in improved yield. More interesting is the extension that found that when the sodium acetate was omitted the reaction led to a set of new complexes that were even more active. Laboratory-scale preparation of enantioselective ruthenium hydrogenation catalysts using standard but specific synthesis techniques shows a wide variation in techniques.

The use of an electrochemical process to generate enantioselective catalysts has been reported, and it is claimed that the catalysts behave differently and give enhanced enantioselectivity. The process appears to have been developed or used in commercial production.

The first example of an enantioselective heterogeneous catalyst is the work of Akabori using a catalyst prepared by impregnating silk with $PdCl_2$ as a hydrogenation catalyst. It was later found that the enantioselectivity was due to the adsorption of optically active amino acid on the palladium. This led to the investigation of a large family of metallic catalysts modified with an optically active additive. However, only two systems have shown sufficient promise. Nickel catalysts modified with tartaric acid give enantioselective hydrogenations of prochiral ketones as β-keto esters and β-diketones. Platinum catalysts modified with cinchona alkaloids show enantioselective hydrogen of keto esters.

The preparation and reaction of tartaric acid-modified active nickel catalysts, as Raney nickel, has been extensively investigated by T. Harada and colleagues and has also been reviewed elsewhere. The simple catalyst preparation involving the use of an activated nickel catalyst with a solution of optically active tartaric acid is enticing, but the number of reaction variables of temperature, pressure, reaction medium, nickel catalyst genealogy, and reaction modifiers is formidable. The system is reported to have been used for the large-scale preparation of optically active 3-hydroxyalkanoic acids and 2-alkanols from the corresponding prochiral ketones. As the capstone of 10 years of research, the recycle of the catalyst, promoted with 2-butylamine, for up to 10 cycles with only a slight decrease in enantioselectivity was reported. A number of optically active additives have been tested, but tartaric acid appears to be the most efficient. It is interesting to note that the substitution of an optically active α-amino acid gives the opposite isomer from one produced by the corresponding α-hydroxyacid.

A study by Holk and Sachtler investigated the use of tartaric acid. They reported that the enantioselectivity of nickel supported on silica, unsupported nickel, and Raney nickel were roughly equivalent but that nickel on alumina was significantly lower. Their study showed that the addition of nickel tartrate to an active nickel catalyst at room temperature enhanced the enantioselectivity compared to the nickel-tartaric acid system, but at 100°C the addition of nickel tartrate to the nickel catalyst gave a lower optical yield than the standard addition of tartaric acid. Without metallic nickel the nickel tartrate gave no hydrogenation. Holk and Sachtler formulated a scheme that requires the presence of nickel both as a tartrate complex and as metallic nickel. The hydrogen activated by the metallic nickel is transferred to the nickel tartaric acid complex so that an optically active hydrogenation reagent is formed in a catalytic system.

Subsequent studies by Tai and colleagues showed that the sonication of the Raney nickel–tartaric acid–sodium bromide system enhanced the enantioselectivity of the catalyst as well as its activity. Their suggested explanation is that the preactivation sonication preferentially destroys the disordered nickel domain, which was not capable of being modified to enantioselective sites and, if not removed, produced a competitive racemate.

The enantioselective hydrogenation of α-keto esters has been investigated by Orito et al. Under optimized conditions, enantiomeric excess in the range of 90% can be obtained. The effect of reaction conditions on activity and selectivity has been studied in detail, and recently the effect of particles has been the subject of investigation. There is still controversy about the relative importance of some of the reaction condition variables in the hydrogenation of pyruvate esters to the lactate esters. It is hoped that recent work by Wells et al. using a standardized platinum or silica catalyst (Europt-1) will set a pattern to reduce some of the uncertainties.

The synthesis of (R)-4-phenyl-2-hydroxybutyric acid as an intermediate for the production of benazepril, an angiotensin enzyme inhibitor, is an example of heterogeneous enantioselective catalysis. The synthesis using a platinum on aluminum catalyst modified with 10,11-dihydrocinchonidine is reported to have been scaled to the 200-kg range for commercial production. Investigation of the reaction variables showed that platinum is superior to rhodium while palladium and ruthenium catalysts give low activity and selectivity. Several closely related modifiers were tested and showed only limited enantioselectivity. Variation in solvent composition appeared to have a minor effect. The modifier concentration was considered with a "ligand-accelerated" reaction and an equilibrium sorption of modifier on the platinum surface.

The preparation of enantioselective heterogeneous catalyst by the immobilization of an enantioselective organometallic complex on a solid substrate has been the subject of much research and discussion with limited success. Recent work by Pugin reports on the immobilization of cationic rhodium complexes. He reacts 3-isocyanatotriethoxysilane with an optically active phosphine carrying an active hydrogen to form the triethoxysilane carbamate of the phosphine. The carefully dried silica gel substrate is then stirred with toluene solution of the carbamate for varying lengths of time and at various temperatures, and the carbamate is locked to the silica by reaction of surface hydroxyl and the ethyloxysilane function. After careful washing to remove unreacted ligand, the material is reacted with 0.8 equivalent of $[Rh(COD)_2]$ BF_4 in methanol until the solution is colorless, and then the product is hydrogen reduced at 1 atm hydrogen. The hydrogenation of methylacetamide cinnamate at 1 atm hydrogen using a ratio of cinnamate to catalyst of 200, showed rates and enantioselectivity in the range of the homogeneous rhodium catalyst. Recycling of the catalyst showed that the rate and selectivity were maintained even though analysis of the first reaction product indicated about a 5% loss of rhodium from the catalyst. The rhodium loss decreased to less than 1% in subsequent cycles. As expected, the efficiency of the catalyst varied with the source of silica gel and even with lots from the same source. It is gratifying to see catalyst immobilization approaching the attributes of a homogeneous catalyst.

A possible hydrid enantioselective hydrogen has been studied by Tungler et al. They investigated the enantioselective hydrogenation of acetophenone by a palladium on carbon catalyst in the presence of (S)-proline. Their study of the kinetics of the hydrogenation showed that the optically active product from acetophenone is the result of the hydrogenation of the condensation produce of acetophenone and proline. The enantiomeric excess increased and the rate of hydrogenation decreased as the ratio of proline to substrate increased, and both decreased as the degree of conversion decreased. The enantiomeric excess was low due to competitive hydrogenation of acetophenone. The reaction should be considered diastereoselective as the asymmetry rises from the formation of the adduct of acetophenone and proline followed by hydrogenolysis to give the optically active product and regenerate proline.

As part of a continuing series of papers on the preparation of "molecular footprint" catalysts, K. Morihara et al. report the preparation of a chiral template catalyst. The catalyst preparation involved the activation of silica gel with hydrochloric acid followed by mixing with a mixture of aluminum chloride and bis(N-benzylcarbonyl-L-alanyl)amine as the template. The product was subsequently dried and used as a catalyst. At this time there is some evidence for chiral activity but not of commercial significance.

CONCLUSION

As stated throughout this chapter, the selectivity of enantioselective catalysts is dependent on many factors of catalyst composition, preparation procedures, and use procedures as well as on many other factors. These must be determined by many trials and analyses. Preparations are discussed in the references and are unique for each product.

27
Polymerization Catalysts

There are many types of polymerization catalysts; most, however, are homogeneous. The few that are heterogeneous, and there is some argument as to whether they are really heterogeneous, will, however, be described. The first is the so-called Phillips polymerization catalyst which is chromia on silica, whereas the second is a titanium or vanadium subchloride prepared by reduction with an aluminum alkyl or chlor alkyl. Both preparative procedures will be briefly described.

The Phillips catalysts are prepared basically by the following general procedure but many variations are possible, some of which will later be described.

Step 1. A particulate 90% SiO_2 and 10% Al_2O_3 mixed oxide having a surface area of at least 400 m^2/g is calcined at 650–750°C (Fig. 24, 25, or 26) in a flow of 90% air and 10% steam to decrease the surface area and increase the pore dimensions.

Step 2. The heat-treated silica alumina is impregnated uniformly (Fig. 37) with an aqueous solution of CrO_3 of such volume and concentration that a content of 3% chromium trioxide is uniformly deposited on the support.

Step 3. The uniform product of step 2 is dried at 150°C and then calcined in air at 450°C for 3 hr. The catalyst is now ready for use.

The silica alumina can be used as microspheres instead of as particulate granules. Also, instead of the chromium being added as CrO_3, it can be added as chromium nitrate, formate, or acetylacetonate, for example, with the heat treatment at 650–750°C in air or oxygen providing the necessary hexavalent chromium.

Instead of the silica alumina, it is possible to use silica, titania, silica-titania, zirconia, or a specially prepared silica aerogel.

There are indications also in the literature that there may be an advantage in titania-modified supports if the catalyst is heat treated in air at 650–750°C, then partially reduced with carbon monoxide, and finally "lightly" oxidized at 300–400°C in air.

ZIEGLER-NATTA POLYMERIZATION CATALYST

It is with some personal concern that this heading is used, because there is much contention as to the priority of the inventions and developments in this area. However, the identity given is descriptive to many interested in this subject, so this heading is used although many other names could very likely be affixed with equivalent accuracy. This matter is for the courts to decide, and it may have been decided by the time this book goes to press. The catalyst itself is referred to as homogeneous, and this too is not fully agreed to by those who contend that it is colloidal or particulate (even though particulate size may be unit crystal size).

A brief description of this polymerization catalyst's preparation follows.

Step 1. Place in a reactor of the type shown in Fig. 9 a quantity of cyclohexane in which is dissolved 1% $TiCl_4$.
Notes: (1) All H_2O, CO_2, H_2S, CO, or any other "chain terminator" must be avoided. (2) The aluminum triisobutyl next to be described is extremely pyrophoric and must be handled with extreme care.

Step 2. While agitating and cooling (hold at 40–60°C), add a 10% solution of aluminum triisobutyl (available from Stauffer Chemical Co.) to produce a ratio of 10 Al to 1 Ti. This is the catalyst solution slurry and is a stock solution from which a quantity is withdrawn for the polymerization.

Instead of the $TiCl_4$ (tickle) above, one can use a hexane solution in which various ratios of $TiCl_4$ or $TiCl_3$ to VCl_3 are used. Varying properties of the resultant polymer are thereby achieved.

Catalyst Regeneration or Metals Recovery

The quantity of catalyst that is used in the polymerization is so small that no desirable properties are lost or undesirable properties caused by leaving the catalyst in the final polymer or copolymer, so there is no "spent catalyst."

28
Dehydration Catalysts

There are many types of dehydration catalysts, usually in the family of oxides such as alumina or silica or mixed oxides such as silica alumina. However, dehydrated calcium or barium sulfate or even dehydrated clays are useful. The preparation of the silica alumina and alumina has previously been described, so the topic will be confined here to a description of the preparation of aluminum phosphate and boron phosphate, both of which are superior in some cases to the typical catalysts previously enumerated. The first one to be described will be aluminum phosphate.

ALUMINUM PHOSPHATE CATALYST

This is a simple catalyst prepared by impregnating gamma-alumina with 85% phosphoric acid solution so as to attain a level of between 10 and 20% P_2O_5 on the alumina.

Step 1. A quantity of activated alumina (i.e., gamma-alumina) in the form of granules or spheres of the preferred size is charged to a pill-coating device such as that shown in Fig. 37.

Step 2. Sufficient solution of 85% phosphoric acid is slowly added to the alumina to give the desired content of P_2O_5 on the alumina (10–20%; lower or slightly greater percentages can also be derived). The reaction between the acid and the alumina is exothermal, so cooling should be provided for the charge in the pill-coating device.

Step 3. The phosphoric acid–coated alumina is either dried by continuing the rotation of the heated bowl of the pill-coating device or by being removed from the pill-coating device and separately dried in an oven such as that depicted in Fig. 22, 25, or 26. After drying, the catalyst is ready for use; however, in some cases it is preferred to calcine it slightly, sat at 300–400°C, in order to completely react the P_2O_5 with the alumina and thus derive a stable catalyst from which the P_2O_5 will not easily sublime.

This catalyst can be modified by varying the quantity of phosphoric acid used, or instead of using alumina for the support of the alumina phosphate one can use silica, silica alumina, titania, or zirconia. These have specific and interesting catalytic properties, which must be examined for the catalytic operation in question.

BORON PHOSPHATE CATALYST

This catalyst is made by mixing H_3PO_4 as an 85% solution with boric acid (H_3BO_3) in sufficient ratio that a 10% excess of P_2O_5 is present. It is conceivable that the support is boron phosphate. The catalyst is a very mild and specific dehydration catalyst that can be used at temperatures up to approximately 400°C. Above that temperature, P_2O_5 will tend to sublime from the catalyst and thus deactivate it.

Step 1. A quantity of 85% phosphoric acid is charged to a kneader-mixer of the type shown in Fig. 27. The temperature is adjusted to 35 °C by passing either cooling or heating water through the jacket of the kneader.

Step 2. Sufficient boric acid (H_3BO_3) is slowly sifted into the phosphoric acid, which still is maintained at 35 °C, to attain a ratio of phosphorus to boron of 1.1:1, that is, with 10% excess of P_2O_5.

Step 3. When the boric acid powder has been completely sifted into the phosphoric acid, a heavy greaselike paste results. This paste is dried at 125 °C (Fig. 22, 25, or 26) and then finally calcined at 350 °C for 3 hr after the temperature of the catalyst has reached this level.

Step 4. The catalyst is cooled to room temperature, with care taken to avoid excessive exposure to atmospheric humidity, which may make the catalyst become tacky on the surface. When it is cool and dry and still brittle, the catalyst is crushed to 100% through a 4 mesh screen using equipment of the type shown in Fig. 29.

Step 5. The crushed catalyst is screened to the size needed for the reaction by passing it through a screening device of the type shown in Fig. 36. After screening, the catalyst should be stored in airtight cans to prevent the reabsorption of moisture.

As for the alumina phosphate catalyst, this catalyst can be varied by varying the quantity of phosphoric acid.

REGENERATION OR RECOVERY OF METALS FROM SPENT ALUMINUM PHOSPHATE AND BORON PHOSPHATE CATALYST

These catalysts are both strongly adsorptive and consequently may, depending on the process in which they are used, be contaminated with toxic materials, so they must be handled with extreme care. The value of components is very low, so recovery is not practiced.

Disposal must be trusted to professionals. Since the boron, phosphate, and alumina are either beneficial or harmless in fertilizers, it is possible that the spent material could find an outlet in this service. The boron and phosphate would both be "slow-release" components, which could be beneficial in fertilizers.

29

The Claus Process
Catalysts

The Claus process is that process whereby hydrogen sulfide and sulfur dioxide are reacted stoichiometrically to produce elemental sulfur and water. The reaction is conducted at a relatively high temperature to keep the sulfur in vapor form. After reaction the gases are passed through a condenser to deposit the sulfur for recovery. The catalyst usually used for this operation is purchased as heat-treated gamma- or alpha-alumina or alumina modified by the incorporation of titania, zirconia, or silica. To modify the alumina, one can impregnate gamma-alumina with titanium tetrachloride (or a Tyzor), zirconium nitrate, colloidal silica (Ludox SM), or one or more of various other colloidal dispersions. After the alumina has been coated or impregnated with the promoters, the impregnated material is dried and then finally calcined at 550–700°C to effect a solid-state reaction between the alumina and the promoter.

DISPOSAL OR RECOVERY OF USED METAL

The ingredients are of low value, so recovery is impractical. Disposal should be either via a procedure agreed to by EPA or foreign equivalent or via professional waste management agencies.

30

Preparation of Automotive Exhaust Catalysts

Automotive exhaust catalysts fall into the large family of total oxidation catalysts. As is true for the entire family of catalysts, there are many physical forms and many chemical compositions, each of which would be most effective under specific conditions. It is generally true that in the abatement of waste streams, pressure drop is of the utmost importance. As a result, the design of the catalyst structure and bed is an engineering problem that must be very carefully considered. A discussion and full consideration of this phase of abatement would no doubt be of interest, but the space required for such a discussion is not tolerable within the plan of this report. As a result, we only point out the fact that some catalysts will be supported on granular supports, others on spherical supports, whereas still others will be coated onto so-called honeycomb structures. All of these types are available from various manufacturers that are well known and are listed in *Chemical Week's Buyer's Guide.*

253

In abatement processes the catalysts are frequently subjected to much more severe and less carefully controlled conditions than apply to a catalyst used in industrial processes. For example, a catalyst may be called upon to initiate the reaction at very low temperatures yet be suitable for operating at extremely high temperatures and after having operated at the extremely high temperatures still be capable of initiating the reaction at low temperatures.

If carbon monoxide is present in the exhaust gas, a typical component of the catalyst is manganese oxide. Manganese oxide is effective, usually at or near room temperature, for the oxidation of carbon monoxide and provides a solution to the low-temperature light-off. If the exhaust gases contain other constituents as well as the carbon monoxide, the CO can be used to raise the temperature of the gas stream to a sufficiently high level that other catalyst components will be effective for the other gas components. Components such as acetic acid, benzene, and methane, which are extremely difficult to oxidize, usually require a temperature, even with a catalyst present, in excess of 400°C. Catalysts that are effective for the olefins, hydrocarbons, ethers, esters, ketones, and alcohols usually are composed of a metal from group VIII as well as copper, chromium, and the rare earths, particularly cerium, all as oxides.

In order for the catalyst to retain its activity at low temperatures and be resistant to deactivation at high temperatures, a so-called refractory oxide is generally intimately mixed with the metals previously mentioned. These oxides can be intimately mixed either by grinding or, preferably, by coprecipitation or a combination of the two. The foregoing is illustrated in the following two general methods of preparation of oxidation catalysts

PREPARATION OF CATALYSTS SUPPORTED ON ALUMINA SPHERES

The following instructions are based on specific conditions, but it will be seen that modifications can be made in all segments of the preparation to achieve conditions most suitable for a different process requirement.

Step 1. Heat 100 g of ⅛-in. alumina spheres having a surface area of 200 m²/g and a total pore volume of 0.4 mL/g to 200°C to remove any adsorbed moisture.

Step 2. Determine the total pore volume of the spheres by taking a sample, heating it to 200°C, cooling and weighing 10 g in a moisture-free environment, then immersing it in water until no bubbles evolve from the spheres. This means that all the pores have been filled with water. Next, dry the spheres by rubbing them between two layers of towels or, preferably, between two layers of filter paper until the surface has been dried to the point where it is just glossy from surface moisture. Weigh the spheres, and determine the increase in weight, which automatically gives the porosity of the spheres.

Step 3. Make a 5% solution of sodium carbonate having a total volume equal to the volume of water determined in step 2 multiplied by the weight of catalyst obtained in step 1.

Step 4. Immerse the catalyst obtained in step 1 in the 5% sodium carbonate solution prepared in step 3. At this point, all of the solution should be completely absorbed by the spheres.

Step 5. Dry the spheres slowly at 200°C long enough to remove all moisture.

Step 6. Prepare a solution having enough platinum chloride to achieve a 0.03% platinum loading on the spheres and of large enough volume to achieve a moist condition with no excess liquid. This can be accomplished by making the solution up to the same volume as the sodium carbonate solution, which in turn is exactly equal to the total pore volume of the sample prepared in steps 1-4.

Step 7. Place the alkalized spheres in a rotating drum such as one of the types shown for pill coating in Fig. 37 D or E. This type of equipment can be obtained from essentially all laboratory supply houses or specifically from A.E. Aubin Co., P.O. Box 1137, Manchester, CT 06040

Step 8. Begin the rotation of the spheres in the pill-coating
device and then add very carefully over a period of 1–2
min the solution of platinum chloride. Continue rotation
until the catalyst becomes uniformly dark from the precipitation of the platinum on the alkalized surface of the
alumina.
Step 9. Remove the coated spheres from the pill-coating machine and immerse them in distilled water so that the alkali can be washed away. Continue washing until a chloride
test indicates that the chloride ion has been essentially
leached out of the spheres.
Step 10. Dry the spheres at 125–150°C. The dried spheres
are the active finished product for oxidation reactions. If,
however, the catalyst is to be used for hydrogenations, a
supplementary reduction can be performed by passing
hydrogen over the spheres at temperatures of 50–200°C.

This completes the preparation of platinum on alumina
catalyst of the simplest type. Palladium and rhodium can be
added to or substituted for the platinum to achieve certain
objectives. Cerium can be coprecipitated with the platinum
or other precious metal to improve the stability of the catalytic metals.

The catalyst, as prepared, has the precious metals primarily on the surface and as such utilizes them most effectively.
However, if there is abrasion of the catalyst, then the precious
metals are most easily removed when the catalytic material
is confined to the surface. The degree of penetration of the
precious metals can be controlled by the quantity of alkali impregnated into the spheres or by eliminating the alkali treatment entirely.

The foregoing describes a catalyst containing approximately 300 ppm of platinum. As the EPA restrictions have
tightened recently, the content of platinum has been increased
to approximately 1000 ppm and rhodium has been added and
then increased to a level of 100 ± 40 ppm. Inasmuch as platinum hovers between $325 and $500 per troy ounce, the cost

of the catalyst has risen, and efforts have been made to substitute some palladium for the platinum; palladium is priced in the range of $80–125 per troy ounce. Rhodium is still another story. It is essential to NO_x abatement and is normally priced at $1000–1500 per troy ounce, but the price has vaulted to as much as $8000 per troy ounce.

It should be noted that ruthenium at $60–80 per troy ounce also is effective for NO_x abatement but becomes a toxic and volatile oxide in an oxidizing atmosphere, which may be encountered. It has not found use in the automotive abatement catalysts.

COATING OF HONEYCOMB CATALYST STRUCTURE

Honeycomb structures are used in those situations where pressure drop is most critical. The honeycomb structures usually can be so engineered that they have a minimum of pressure drop because there is a minimum of turbulence in the bed. Honeycomb structures are built by a number of different suppliers listed in *Chemical Week's Buyer's Guide.*

Step 1. A honeycomb structure 8 in. wide, 15 in. long, and 2 in. thick is immersed in a slurry made by dissolving 300 g of aluminum nitrate nanohydrate in 4 liters of distilled water to which has been added 500 g of finely divided alumina hydrate. The solution and solids are milled in a ball mill for 3 hr with 1 liter of ½-in fused alumina balls as grinding medium.

Step 2. The support coated with the wet slurry is now blown gently with an air jet to remove excess slurry.

Step 3. The wet coated honeycomb is dried at 150°C and finally calcined at 400°C for 2 hr to convert the aluminum nitrate to aluminum oxide. The purpose of the aluminum nitrate is to act as an adhesive for the alumina hydrate, which otherwise would make a very powdery, chalky film.

Step 4. The honeycomb, now coated with a thin adhesive film of aluminum oxide, can again be coated by a second immersion as stipulated in step 1. Usually the honeycomb is immersed upside-down with respect to its previous immersion (with its former top now at the bottom), so that drainage can produce a more uniform film from end to end of the openings. The coating as described produces the alumina-type "wash coat," which is one of the most popular and frequently used. However, mixed rare earths, cerium oxide, magnesium oxide, or chromium oxide, for example, can be used instead of alumina or together with the alumina. In the case of magnesia, the product would, of course, on calcining become magnesium aluminate spinel, which has very desirable characteristics differing from those of alumina. Zirconia can also be added for various benefits for certain types of abatement reactions.

Step 5. Two liters of solution is prepared having a concentration of 1% platinum as chloroplatinic acid. The pH is adjusted with ammonium hydroxide or sodium carbonate to a pH at which incipient precipitation of the platinum begins to occur (ca pH 3.5).

Step 6. The alumina-coated honeycomb is immersed in the platinum chloride solution for a period of 30 sec to 1 min.

Step 7. The moist platinum-coated honeycomb is dried and calcined at 150°C. After the first impregnation, a second impregnation, as specified above, can be effected together with redrying and recalcining.

Step 8. It may be desirable to remove the chloride ion, and this can be done by washing it with water or reducing it in hydrogen. If, however, the catalyst is to be used at high temperatures, it is not necessary to remove the halide in a separate operation as it will generally be removed during use. However, if the chloride is present as a metal alkali, it must be removed by washing or ion exchange before use at 600°C or above.

Step 9. Instead of the precious metals, one can, of course, use the base metals such as manganese, nickel, cobalt, and copper, or even iron although iron is generally one of the weaker of the oxidation catalysts. These are generally impregnated onto the support as nitrates, but in some cases acetates or complex ammines can be used advantageously. As stated above, the stability of the catalyst can be improved even when it is supported on the alumina film by adding a costabilizer such as the rare earth group or individual members of that group, or other refractory oxides (thoria, magnesia, and the like). Thoria is generally rejected because it would be radioactive although at a very low level.

Note: The catalyst just described is useful for a broad range of total oxidations as well as for automotive fume abatement. The structure must, of course, be of specific dimensions, but the "wash coat" is the same; together with the platinum at about 800 ppm, rhodium at 60–150 ppm and palladium are acceptable as a substitute for some platinum.

REGENERATION OF PRECIOUS METALS CATALYST AND METALS RECOVERY FROM ALL TYPES OF PRECIOUS METALS CATALYSTS

If the catalyst has been deactivated by carbonaceous deposition, it can generally be regenerated by a careful oxidation to remove the carbonaceous material that is acting as a physical poison. Poisoning by sulfur or halides can also be overcome by oxidation at temperatures exceeding 300°C, preferably above 500°C.

METALS RECOVERY

Catalysts that do not respond to the foregoing or other regeneration techniques are subjected to metals recovery, which can be accomplished in one of the following procedures.

Acid Dissolution

An acid extraction procedure can be performed if the metals on refractory are accessible to the acid. The catalyst is crushed by grinding or ball milling to produce a fine powder. This powder is treated with aqua regia (3 HCl + HNO_3) to produce chlorine, which converts the metals to water-soluble chlorides. In the case of carbonaceous coating, this is carefully oxidized, leaving the metals as an ash that is treated with aqua regia to dissolve the precious metals and recover the water-soluble chlorides.

An additional "wet" method of metals extraction and recovery has been studied and reported from the Department of Mining and Metallurgical Engineering at Montreal, Quebec. This process is not reported to have been commercialized.

The process, briefly, starts with the reduction of the spent catalyst to reduce the oxides, primarily those of rhodium, which are acid-insoluble. The solution used to dissolve the Pt and Rh is composed of HCl at various concentrations and nitric acid, also of various concentrations, along with various quantities of aluminum chloride. The temperature of extraction is either at the boiling point of the acid solution or at an elevated pressure and correspondingly high temperature.

At elevated pressure the extraction time can be reduced to 30 min versus the 2 hr required at atmospheric pressures.

The best extraction recovery is 95% of the platinum and 82% of the rhodium.

Recovery by Fusion and Dissolving

There are several procedures that involve fusion of the spent catalyst. First to be mentioned is an earlier procedure in which the precious metals on alumina were fused with an alkali hydroxide or carbonate, rendering the alumina soluble as sodium aluminate and leaving the precious as an unaltered solid. The metal is removed by filtration and thus recovered.

The disposition of the sodium aluminate posed an economic problem for this process.

A patented process practices by Multimetco, Inc., of Anniston, Alabama, consists in melting the refractory support and metals in a plasma arc. The metal, because of its density, migrates toward the bottom of the melt and into a concentrate, which is processed as a high concentration ore.

The U.S. Bureau of Mines has developed, through pilot-plant scale, a process in which the pulverized metals-containing spent catalyst or ore is treated with a cyanide solution at 100°C or above to dissolve the metals. The waste solids are filtered off, and the cyanide solution containing the metals is heated to decompose it and precipitate the metals. This is described and claimed in U.S. patent 5,160,711.

31

SO_x and NO_x Abatement Catalysts

CATALYSTS FOR THE ABATEMENT OF NO_x AND FOR THE SIMULTANEOUS ABATEMENT OF $SO_x NO_x$

The removal of NO_x from exhaust gases also containing oxygen is accomplished in two basic ways. One process, identified as selective catalytic reduction (SCR), involves the selective reduction of the NO_x using ammonia as the reducing agent. Currently used commercially in Japan, this process was first developed in the United States.

The second type of method relies on the selective adsorption of the NO_x and its subsequent release in a concentrated form in an O_2-free gas flow, which then passes together with a reducing gas such as H_2, CO, or methane over a reduction catalyst. NO_x is removed to an extent of 95+%. Several processes of this type are in various stages of development.

The first of these processes is the NOXSO process, which briefly comprises the fluidized adsorption of the NO_x and SO_x with an alkali on alumina adsorbent. These adsorbates are regenerated in a concentrated form, with the NO_x reduced to N_2 and water and the SO_x to H_2S, which is reduced to sulfur in the Claus process.

The second is in the developmental stage but has passed the semiworks in a side-stream reactor at a steam-electric generating station. Negotiations for a demonstration plant are under way. It is identified as the COL-UD process.

The third is suitable only for NO_x abatement and involves a uniquely effective adsorbent, the partial evolution of the NO_x from the adsorbent coupled with a reduction of most of the NO_x while it is still adsorbed and being evolved. What is evolved is subsequently reduced to H_2O and N_2 as a concentrated stream over a reduction catalyst.

Procedures for the preparation of the adsorbent catalyst and/or catalysts for all four processes follow.

CATALYST FOR THE SELECTIVE REDUCTION OF NO$_x$ IN THE PRESENCE OF OXYGEN

The catalyst used by the SCR process is vanadium pentoxide and titanium dioxide on a support, usually aluminum oxide of the gamma species. Other supports can be used: silica, silica alumina, and chromia. The procedure is as follows:

Step 1. Gamma-alumina in the form of spheres or granules of the desired size and surface area is chosen, and 100 parts is weighed out.

Step 2. The alumina is heated to 250°C for 2 hr after reaching this temperature and then is allowed to cool in an environment in which reabsorption of the moisture can be avoided.

Step 3. An aqueous solution is prepared that has 7% NH$_4$VO$_3$ content. Simultaneously an aqueous solution is prepared of Tyzor LA (DuPont) containing 5% TiO$_2$ on a solid basis. This must be made up daily because the ester of titanium slowly hydrolyzes in water.

Step 4. The alumina granules are immersed in the Tyzor solution, drained, dried, and calcined at 250°C for 2 hr. This step is repeated if necessary to get a TiO$_2$ loading of 5–7% TiO$_2$.

Step 5. With the alumina granules now impregnated with the TiO$_2$, calcined, and cooled, they are immersed in the ammonium vanadate solution, drained, dried, and calcined at 250°C for 2 hr. This is repeated if necessary to obtain a loading of 5–7% V$_2$O$_5$.

Step 6. The catalyst is now ready for use.

Alternative

1. The ingredients can be applied to the granules or spheres by spraying, using equipment such as that depicted in Fig. 37.
2. The catalyst can be derived as a fluidized bed candidate by using microspheroidal alumina.

As part of the SCR process, it may be necessary to separately remove the SO$_2$ in a reactor using an alkali carbonate-impregnated support, usually alumina. This is used either as a fixed bed or as a fluidized reactor. The usual loading of the alumina is 12–20% alkali carbonate.

PREPARATION OF CATALYST FOR THE NOXSO PROCESS

This catalyst is actually a chemisorbent composed of an alkali carbonate impregnated into granular, spherical, or microspheroidal alumina. The process is a simple one, and the adsorbent process is identical to that described in the immediately preceding section.

A NEWLY DEVELOPED PROCESS AND CATALYST ADSORBENT CYCLING BETWEEN ADSORPTION AND REGENERATION

This procedure involves the application of catalyst in a reducing atmosphere in which the NO_x is reduced to N_2 and H_2O and the SO_x is reduced to either hydrogen sulfide or elemental sulfur. The catalyst is made by the following detailed procedure.

Step 1. Make an aqueous solution comprising a total volume of 2 liters in which are dissolved 375 g of aluminum nitrate nonahydrate adn 291 g of cobalt nitrate hexahydrate (Fig. 8).

Step 2. Adjust the volume to exactly 2 liters, and heat the solution to 30°C. Have the precipitation equipment set up in such a way that the nitrate solution can be agitated and concentrated ammonium carbonate solution and carbon dioxide gas can be added continuously during both the precipitation and digestion period. CO_2 flow rate should be 250 mL/min.

Step 3. Separately prepare a concentrated solution of ammonium carbonate, and place this in a separatory funnel so it is just above the beaker in which the nitrate solution is being agitated.

Step 4. With the temperature at 30°C and the agitation being carried out vigorously, add the concentrated ammonium carbonate solution to the solution of cobalt and aluminum nitrates until a pH of 6.8 ± 0.2 has been reached.

Step 5. At the conclusion of the precipitation, continue agitation at 30°C for an additional hour, during which carbon dioxide is being continuously added.

Step 6. Filter the precipitate, discard the filtrate, and dry the filter cake for 24 hr at 140 ± 25°C.

Step 7. Calcine the catalyst in suitable containers, preferably stainless steel or unglazed porcelain, at 350°C for 2 hr after the catalyst reaches that temperature.

Step 8. To a 2-liter ball mill containing 500 mL of ½-inch grinding pellets, add the 75 g of powder produced in step 7.

Step 9. Add to the ball mill 100 g of colloidal ceria as a 22% solution of the type produced and sold by Rhone-Poulenc.

Step 10. Finally, add to the ball mill 325 mL of distilled water.

Step 11. Mill this mixture of powder, water, and colloidal ceria for a period of at least 4 hr, preferably for 12–16 hr.

Step 12. Remove the slurry from the ball mill and retain it in a closed vessel.

Step 13. Put 200 mL of the slurry after the ball milling into a 500-mL beaker. Immerse one of the 2-in.-long by ½-in.-diameter honeycombs in the slurry and remove after the slurry has completely coated the surface.

Step 14. Using a hair blower with the heat on "high," blow the surface of the honeycomb until the film that is adhering to it has been dried. Repeat steps 13 and 14 once or twice or until the honeycomb is uniformly and completely coated.

Step 15. Calcine the coated honeycomb at 350°C for 3 hr. Catalyst pickup should equal 50% of the initial weight of the honeycomb. After the final calcining, the catalyst is ready for use.

FUME AND ODOR ABATEMENT CATALYSTS IN GENERAL

There is really no typical industrial discharge that can be used to illustrate a general or model procedure. Frequently the effluents comprise fumes, odors, combustibles, and organic-inorganic aqueous discharge situations. Various conditions may be encountered:

A. Combustibles in high and constant concentrations—unusual
B. Combustibles in low and variable concentrations
C. Refractory combustibles requiring, even with catalytic assistance, high initiation temperatures
D. Odors alone in very low concentrations—will not independently support combustion
E. Effluents containing sulfur compounds
F. Effluents containing a halide as an organic compound or as HCl and the like
G. Effluents as aqueous solutions containing organics, inorganic salts, and an alkali salt of an organic acid

POSSIBLE TREATMENT PROCESS AND CATALYSTS FOR FOREGOING SITUATIONS

In situation A, there are a large number of options as to both catalyst and process. Typically, however, combustion occurs in a single- or multiple-tray reactor with heat recovery from the effluent and preheating of the inlet gas. Catalysts that are useful are copper chromite or precious metals on a refractory oxide in the form of granules, cylinders, rings, or honeycomb. The unit must be protected from overheating by excessive fuel or inorganic solids and from an explosive mixture with air. Descriptions of the catalyst preparation are given in Chapters 18 and 24 under copper chromite and precious metals on alumina and honeycomb.

In situation B, the same catalysts can be employed, but usually preheating must be provided as well as means of controlling the catalyst temperature by controlling preheating and dilution of the inlet gas by air or inert gas.

In case C, the catalyst usually is a supported precious metal, and the inlet gas must be preheated or a supplementary fuel added.

In case D, the odorous materials are at a high enough temperature for combustion, and the catalysts are either copper chromite or the supported precious metals. If the gas is cold it may be necessary to first adsorb the odors and then evolve them as a concentrated mixture with air, heat this much smaller volume, and pass it over one of the aforementioned catalysts.

In case E, following combustion over a supported precious metal catalyst, the SO$_2$ is adsorbed or is oxidized to SO$_2$ for acid synthesis.

In case F, the combustion is again carried out over a supported precious metal catalyst, and the effluent is scrubbed by an alkali solution to adsorb the acid gases NO$_x$, SO$_x$, and HC1 or other halide.

In case G, it may be necessary to strip out the organics and oxidize them while the aqueous solution of inorganic salts is disposed of by sewering or in another environmentally acceptable manner.

Appendix: Catalyst Metals Values Reclaimers

ACI Industries, Dublin, OH and Brussels, Belgium   Recl.
Purchases spent base and noble metal catalysts; has contacts with metal refiners. About 80% of its business is in spent catalysts, all of which is for reclamation (no warehousing over extended periods of time and no landfill). Was being recommended by Harshaw in USA and Harshaw Chemie in Europe in attempt to offer a full catalyst service. Handles spent nickel catalyst in fatty acids and as Raney nickel for Tomah, Milton, WI and Pedricktown, NJ. Also does business with Exxon Company USA (Elton Forbes). (Used to be: Falconbridge Trading Associates.)
   USA: Vincent Campbell (614) 764-8566; FAX (614) 764-7803
   Europe: Scott H. Fisher 32-2-536-8646/8647; FAX 32-2-536-8600

AGMET Metals Inc., Solon, OH   Recl.
Specializes on industrial by-products, such as catalysts, filter cakes, and sludges containing heavy metals such as Ni, Co, Mo, W, and Cu. Also able to handle V and Zn. Has processing arrangements for re-

clamation. A new recycling plant is under construction at Solon, OH; plant will produce various ore supplements. Will issue certificate of reclamation.

Dennis J. Dunagan (216) 439-7400; FAX (216) 739-7446

AMC (Allied Metals/American Processing), Detroit, MI   Recl.
Capable of handling spent catalysts and any related metal-bearing streams containing Ni, Co, Mo, W, Cu, Zn, and others. In case of precious metals, offers either outright purchase or the return of the precious metal content (Ni = 10% min; P = 0.2% max; S preferably less than 0.5%). Spent catalyst material is sent to offshore facility to be treated; metals and any energy values are extracted, and remaining by-products are disposed safely. In two instances customers requested letters of reclamation. Four requests were received from Exxon, but no business was awarded (as of end of 1989).

Jennie Thomas (313) 368-7110; FAX (313) 368-2107

Amlon Metals Inc., New York, NY, England,   Recl.
and Other Countries
Purchaser of spent catalysts (Ni, Co, Cu/Zn, Pd, Pt, Mo, etc.), chemicals, and other by-products. Intermediate plant handling spent catalysts (drying, blending, and calcining) in England. Other offices in many European countries and in Brazil. Has grown considerably in recent years. Issues certificate of consumption. No landfill.

Lee Lasher (212) 742-1043; FAX (212) 227-4028

BPM Industries, Inc., Schererville, IN
See its subsidiary, Resource Chemical Company.

Bullard Feed Company, Bremen, IN   Recl.
Interested in HTS and LTS spent catalysts, but only if dry, according to UCI.

Mike Sandroni (219) 546-5560
Refer to: Norchem Industries Inc.

Cal-Chem Metals Inc., Huntington Beach, CA
Don Nickerson (714) 962-6640; see Oatfield Industries Inc.

Catalyst Disposal Services (Canada) Limited,   Recl.
Calgary, Alberta, Canada
See IBR (Industrial Byproduct Recycling Inc.)

CALCORP Resources Inc., Houston, TX   Recl.
Purchases spent catalyst for reclaimed value of metals contained, such as low-grade Pd from acetylene converter catalyst. Has contacts with metallurgical companies worldwide for reclamation of metals contained. No landfill! Can issue certificates of destruction, etc. Only handles nonhazardous materials. Has long-standing experience in finding homes for spent ammonia manufacturing catalysts including Cu/Zn LTS, ZnO, HTS, primary and secondary steam reforming, methanation, and FeO ammonia synthesis catalysts. Maintains contacts with Rob Davis, BOP, Jim Burke, BRCP, and Marc Woolston, Sarnia, on spent Pd catalysts.
  John Calpakia (713) 443-4618; FAX (713) 443-4171

Cat-Tech (Catalyst Technology), KY and other Locations   Handling
Recently purchased by Shell, affiliated with CRI. Provides all kind of catalyst-handling services such as unloading and loading. Handling 80% of all catalyst loading/unloading business in USA for ammonia and hydrogen plants including hydroblasting.
  Bill Taylor, VP (domestic) (502) 222-0121; FAX (502) 222-0212 ext. 213
  Robert Allen, Baton Rouge, LA (504) 642-5933

ChemCat Resources (also Kemstar Corporation),   Reselling
Los Angeles, CA
Resells surplus catalysts and chemicals. Resells mildly used catalysts that can be reused or cascaded.
  David Tress (213) 390-0189 and (914) 681-0126; FAX (213) 391-8143

Climax Molybdenum Company, Ann Arbor, MI   Ref.
Manufacturer of molybdenum, molybdophosphoric acid, etc.
  Ann Arbor, MI: Roger Sebenik, Robert A. Ference, C.C. Clark (Amax Tungsten Div.) (313) 481-3000
  Greenwich, CT: G.N. Otwell (Amax Molybdenum Div.) (203) 629-6400.

CPI (CP Industries), Joliet, IL   Ref.
Operates six refining plants in several states. Refines waste treatment sludges; also takes in spent catalysts such as wet (15% Ni) and dry (44% Ni) nickel catalysts. CPI recovering, for metal value, Ni, Cu/Zn, Cu, Cr, Co/Mo. Produces hexavalent chromium.
Address: CPI, Industry Avenue, Jolliet, IL 60435
  Thomas A. Cassata (815) 727-1074; FAX (815) 476-7984

CRI – Catalyst Recovery, Inc., Baltimore, MD   **Reg.**
Recently (1989) purchased by Shell and thus affiliated with Criterion; offers catalyst offsite regeneration at plants in Lafayette, LA; Luxemburg; Alberta, Canada; and Japan. More recently, added "LDG" length and density grading for regenerated catalysts. Probably has not regenerated catalysts used for ammonia/hydrogen manufacture.
Brad Housenga (713) 874-2693; FAX (713) 874-2641

CRI-MET, New Orleans, LA   **Ref.**
Joint venture between Shell and Amax for total recycling of spent hydrotreating and other catalysts. Stated capacity of the Braithwaite, LA plant is 30,000 tons/yr of spent hydrotreating catalyst. Current charges are $200/ton for nonhazardous, and $300/ton for hazardous materials; all credits for metals accrue to CRI-MET. Extra charged apply if silica content of spent catalyst is greater than 4% and if metal content is lower than usual. Can handle tungsten as of December 1989. Four basic products are produced: molybdenum sulfide (for ferromolybdenum alloy manufacture): vanadium pentoxide; Ni and Co concentrate ("NiCo mat," which is used by Amax as raw material); and alumina trihydrate (mold for water treatment applications).
John N. Glover (713) 874-9898; FAX (713) 874-2631

CSI Catalyst Services Inc., Shelbyville, NY
Provides catalyst testing, certification, and consulting services for the hydrogen and ammonia industries.
Robert Habermehl (502) 722-5615

Cypress Miami Mining Corp., Claypool, AR   **Ref.**
Recovers precious metals such as from catalytic reforming and ammonia/nitric acid catalysts. Recycles platinum for Redwater ammonia plant and reclaims precious metals from spent Powerforming catalyst for Esso Petroleum, Canada. Recovers platinum from scrapped heat exchanger and boiler tube metals.
Stuart A. Alperin and Paul Amorosa (617) 599-9000;
FAX (617) 598-4880

Falconbridge International Ltd.,   **Recl.**
subsidiary of Falconbridge Limited
Falconbridge Limited, a major Canadian mining company, is recovering Ni, Cu, Co, Zn, Au, Ag, Pt, Pd, and Rh from residues, spent catalysts, etc., by blending such materials with the product from its

own mines prior to smelting. The Falconbridge Smelter Operations have an agreement with the Canadian government for such activities.
Gaston J. Reymenants 32-2-673-4180; FAX 32-2-660-8545

Frit Industries, Ozark, AL   Recl.
Processes spent catalysts into fertilizers, especially Cu/Zn and ZnO, but no Cr.
Reba Raines (205) 774-2515; FAX (205) 774-9306

Gemini Industries Inc., Santa Ana, CA   Ref.
Recovers noble metals from spent catalytic reforming catalysts. Principle operation is to recover and refine platinum group metals. Alum (aluminum hydroxide solutions) is being made as a coproduct and sold to water treatment plants. Carbon-based catalysts can also be handled. Has been processing spent Pt and Pt/Re reforming catalysts for Exxon Company USA since 1982.
J.P. Rosso (714) 250-4011; FAX (714) 250-6422

Gulf Chemical and Metallurgical, Freeport, TX   Ref.
Owned by Societe Generale de Belgique. Recovers Mo, V, etc. at Freeport, TX. Implements through sister companies total metal reclamation including Ni, Co, and alumina (used for carpet sizing, etc.). Also interested in Cu/Zn, Ni/W, and ZnO. Uses EPA-rated BAT (best available technology), i.e., a pyrometallurgical approach.
Jay Jaffee (409) 233-7882; FAX (409) 233-7171
see comprehensive file

Hall Chemical Company, Arab, AL   Recl.
Robert Fox

IBR (Industrial Byproduct Recycling Inc.),   Recl.
Calgary, Alberta
Actively involved as broker and trader in recycling of industrial byproducts including (but not limited to) catalysts. Has its own R&D capabilities. Has secrecy agreements with many catalyst manufacturers. Provides for recycling of FCC catalyst of many U.S. refiners; spent FCC catalyst is used in cement manufacture. Has developed recycling outlets for high-temperature-shift (iron-chromium) catalyst. Has provided proposals for recycling of copper and arsenic catalysts. Well versed in environmental matters. Only catalyst not able to handle is lead oxide catalyst. Maintains another office in New Providence, NJ.
Lori S. Feldman, Randy Jaggard (403) 264-8778; FAX (403) 266-4427

Jacomij Metalen B.V., Postbus 100, Duurstede, The Netherlands
European source for spent catalyst disposal.

Millmet Resources, Inc., Novi, IL
Recycles brass and zinc, also spent ZnO catalyst.
P.O. Box 246, Novi, MI 48050
  (313) 437-8114; FAX (313) 437-4748 Attention Mardy or Chuck

Norchem Industries, Inc., Bremen, IN   Proc.
Interested in spent catalysts for manufacturing mineral premixes
for both feed and fertilizer use; also has contact with processors. (Interested in HTS and LTS, but only if dry).
P.O. Box 127, Bremen, IN 46506
  Mike Sandroni (219) 546-5560; FAX (219) 546-5517

Oatfield Industries Inc., Mobile, AL   Proc.
Purchases spent catalyst, primarily Cu, Cu/Zn, and Zn. Processes
the spent catalyst and sells it as micronutrients, such as in trace
mineral business for fertilizers. Also handles manganese and chromium
for alloy manufacture. Has 30 acres, used for import/export businesses, located at 3151 Hamilton Road, Theodore, Alabama 36582.
Associated with ALA-MEX imports manganese ore and nickel. Buys
copper by-products.
  Ron Dunn (205) 653-1665 and (205) 653-8700; FAX (205) 653-5340
  See also Cypress Miami Mining Corp.

Parkins International, Inc., Houston, TX   Recl.
Purchases spent base and noble metal catalysts. Has some processing on-site such as a rotary drier and a small still.
  Mary Collins (713) 675-9141; FAX (713) 675-4771

PGP Industries, Inc., Sante Fe Springs, CA   Recl.
Refines precious metals and recovers Pt and Pd.
  Dan Hernandez (213) 926-9489 FAX (213) 926-9489
  Southwest area: Elmer Silmindinger (817) 488-8116

Platina, Plainfield, NJ   Recl.
Reclaims noble metals from spent catalysts. Marketing handled by
Barker Associates.
  Frank Barker (609) 227-4256

Powmet, Rockford, IL
Handles base and noble metal catalysts as traders. Does some processing of its own. Has outlets for alumina. Interested in copper and

especially in ZnO. Working with Exxon on rhenium recovery from catalytic reforming catalysts.

Bill Thiede (815) 398-6900; FAX (815) 398-6907

Resource Chemical Company (RCC), Schererville, IN   **Recl.**
Ron Tenny, its president, is experienced in managing wastes such as wastes from steel industry and spent Cu/Zn, Ni, Ni/W, and phosphoric acid–containing catalysts. Provides shipping service, has trucking permits to handle wastes, and has several sites for accumulation of EPL-approved hazardous waste storage. Pollution test laboratory with 18 people can classify EP toxicity. Will issue certificates of reclamation. Made arrangements with Western Mining (Australia) for recycling of many types of spent nickel-containing catalysts. The spent catalyst is shipped to Australia and added to ore-based smelter feed (flux) to enrich Ni content. Handles all of Chevron's Ni/W hydrocracking catalysts. Works with Royster Agricultural to incorporate spent phosphoric acid on kieselguhr/silica catalysts of UOP and UCI with the phosphate rock feed to the TVA (Tennessee Valley Authority) granulation process for producing NPK fertilizers. Handles spent catalyst of Baton Rouge and Sarnia higher olefin plants and imports spent phosphoric acid (UOP and UCI) catalysts from Europe (Sweden).

Ronald Tenny, President (219) 322-2573; FAX (219) 322-8533

Sabin Metal Corp., New York, NY   **Rec.**
Recovers precious metals using pyrometallic and hydrolytic processes at Rochester, NY.

Les Cline (212) 349-4800; FAX (212) 964-0276

SCA Chemical Waste Services, Inc
See extensive file.

Western Mining Corporation (Westminco), Australia   **Ref.**
Western Mining Corporation Limited of Kalgoorlie in Western Australia produces nickel from ore. Trying to import spent nickel catalysts to enrich the nickel content of the ore charged to the smelter. Actively working with Resource Chemical Company to coordinate shipments of spent Ni catalysts to Australia.

W.H. Cunningham, Mgr. Matte & Byproduct Sales, Belmont 6104, Australia Tel. 6194780871

Don R.T. Hall, Mgr. Nickel Smelter, Kalgoorlie, West Australia
Tel 090 22 0333; FAX 090 21 6240
See Resource Chemical Company.

Others

Lauro's Marketing Lt., Edmonton, Alberta, Canada (403) 484-1437

Pashelinsky & Sons, Jersey City, NJ (nothing below 35% Ni on dry basis).
(201) 333-6606

Texas Glechna Steel Casting Company.

References

Acrylonitrile Production Catalyst. U.S. Patent Application 80/150, 151

Activity Enhancement of High Silica Zeolites. U.S. Patent 4,435,516. C.D. Chang, J.N. Miale. (1984).

Y. Izumi Agnew, *Chem. Int- Ed, Eng. 10*:871 (1971).; M.J. Fish, and D.F. Ollis, *Catal- Dev- Sel. Eng., 18*:259 (1978).

S. Akabori, et al.; *Nature* (London) *178*:323 (1956).

Ammoxidation Catalysts. U.S. Patent Application 219,708.

Ammoxidation of Olefins in the Presence of Multiply Promoted Tin-antimony Oxide Catalyst. U.S. Patent 4,339,394. R.K. Grasselli, C. Falls, D.D. Suresh, J.F. Brazdil, and F.I. Ratka (1982).

R.B. Anderson, Chain Growth and Iron Nitrides in Fischer-Tropsch Synthesis, Dept. of Chemical Engineering, McMaster Univ., Hamilton, Ontario, Canada. A personal copy of a subsequent publication.

Armor, John N., *Applied Catalysis B. Environmental*, 1992, 221–256. "Environmental Catalysis" Review.

Battery Housing Assembly with Integral Limited Travel Guide Rail. U.S. Patent 5,242,767 K.J. Roback and T.E. Derdzinski. (1993).

Biological Process for the Elimination of Sulphur Compounds Present in Gas Mixtures. U.S. Patent 5,236,677 M.D. Torres-Cardona, S. Revah-Moiseev, A. Hinojosa-Martinez, F.J. Paez-Moreno, and V.M. Morales-Baca. (1993).

Bismuth Molybdate on Silica Catalyst. U.S. Patent 3,497,461. W.R. McClellan and A. Stiles (1970).

H.U. Blaser, Heterogeneous Catalysis and Fine Chemicals 111, 1993, Elsevier Science Publisher B.V., New York.

H.U. Blaser et al., *Stud. Surf. Sci. Catal.*, 41:153 (1988).

D.W. Breck, (ed). Zeolite Molecular Sieves, Structure, Chemistry and Use. John Wiley & Sons, New York (1973) pp. 725–731.

Catalyst and a Polymerization Process Employing the Catalyst. U.S. Patent 4,234,710. C.W. Moberly, M.B. Welch, and L.M. Fodor (1980).

Catalyst for the Production of Ethylene Oxide. U.S. Patent 3,962,136. R.P. Nielsen and J.H. Rochelle (1976).

Catalytic Cracking. Improving the Olefinic Gasoline Production of Low Conversion Fluid. U.S. Patent 3,894,931. D.M. Nace and H. Owen (1975).

Chemical Production of Metallic Silver Deposits. U.S. Patent 3,702,259. R.P. Nielsen (1972).

Conversion of Hydrocarbons with Y Faujasite-type catalyst. U.S. Patent 3,894,936. H. Owen (1975).

Conversion of Nitrogen Oxides. U.S. Patent 4,157,375. S.M. Brown, and G.M. Woltermann (1979).

Converting Hydrocarbons with Zeolite ZSM-4. U.S. Patent 3,923,639. J.C. Pitman (1975).

J. Cornier, J.-M. Papa, M. Gubelmann; Industrial Applications of Zeolites; L'Actualite Chimique, Nov.-Dev. 1992; pp. 405–414.

Cracking Catalyst. U.S. Patent 3,912,619. J.S. Magee and R.P. Daugherty (1975).

Cracking Catalyst and Process of Cracking. U.S. Patent 3,252,889. R.G. Capell and W.T. Granquist (1966).

Cracking of High Metals Content Feed Stocks. U.S. Patent 3,944,482. B.R. Mitchell and H.E. Swift (1976).

Crystalline Chromosilicates and Process Uses. U.S. Patent 4,363,718. M.R. Klotz (1982).

Crystalline Metallophosphate Compositions. U.S. Patent 4,310,440. S.T. Wilson, B.M. Lok, E.M. Flanigen (1982).

Crystalline Silica. U.S. Patent 4,061,724. R.W. Grose and E.M. Flanigen (1977).

Crystalline Silicoaluminophosphates. U.S. Patent 4,440,871. B.M. Lock, R.L. Patton, R.T. Gajek, T.R. Cannan, E.M. Flanigen.

Crystalline Zeolite Y. U.S. Patent 3,130,007. D.W. Breck.

Crystalline Zeolite ZSM-11. U.S. Patent 3,709,979. P. Chu.

Crystalline Zeolite ZSM-20 and Method of Preparing Same. U.S. Patent 3,972,983. J. Ciric (1976).

M.E. Davis and S.L. Suck (eds). Selectivity in Catalysis, American Chemical Society. Catalysis and Surface Science Secretariat: Chemical Congress of North America (4th: 1991, New York, NY). ACS Symposium Series 517.

Dysprosium Zeolite Hydrocarbon Conversion Catalyst. U.S. Patent 3,914,383. F.W. Kirsch, D.S. Barmby, and J.D. Potts (1975).

E.L. Force and A.T. Bell, *The Relationship of Adsorbed Species Observed by Infrared Spectroscopy to the Mechanism of Ethylene Oxidation Over Silver. J. of Catalysis, 40*:356–371 (1975).

Heat Treated Bismuth Molybdate and Phosphomolybdate on Titania Catalyst. U.S. Patent 3,640,900. W. McClellan and A. Stiles (1972).

Heterogeneous Catalysis and Fine Chemicals III, Elsevier Science Publishers BV, New York, (1993), p. 99.

High Temperature Shift Catalyst and Process for its Manufacture. U.S. Patent 4,861,745. D. Huang and J. Braoden (1989).

J.P. Hogan, D.D. Norwood, and C.A. Ayres, *Phillips Petroleum Company Loop Reactor Polyethylene Technology*, J. of App. Sci. Applied Science Symposium, 36:49.60 (1981).

J.P. Hogan, Ethylene Polymerication Catalysis Over Chromium Oxide, J. of Poly. Sci., A. *1*, 8:2637–2657 (1900).

P.J. Hogan, A. Stewart, and T.V. Whittam. Synthetic zeolite Nu-10

−useful for sepg. aromatic cpds. etc. and esp. for methylation of toluene. EPO 065,400-1986.

A. Holk and W.M.H. Sachtler, Enantioselectivity of Nickel Catalysts Modified with Tartaric Acid or Nickel Tartarte Complexes. *J. Cat. 58*:276, (1979).

Hydrocarbon Conversion Catalyst and Preparation Thereof. U.S. Patent 3,360,484. S.M. Laurent (1967).

Hydrocarbon Conversion Catalysts. U.S. Patent 3,367,885. J.A. Rabo and P.E. Picket (1968).

Hydrocarbon Conversion Over Activated Erionite. U.S. Patent 3,925,292. W.P. Burgess (1975).

Hydrocracking Process. U.S. Patent 3,360,458. D.A. Young (1967).

Hydrodesulfurization Process With a Portion of the Feed Added Downstream. U.S. Patent 3,923,637. R.D. Christman, J.D. McKinney, T.C. Readal, and S.J. Yanik (1985).

Y. Izumi, *Advances in Catalysis* (D.D. Ely, H. Pines and P.B. Weisz, eds.) Vol. 32, 215, Academic Press, San Diego, 1983. H.U. Blaser and M. Muller, *Stud. Surf. Sci. Catal., 59*:73 (1991).

J. Chem. Soc. Chem. Commun. 7:659–60 (1993).

K.E. Koenig, et al; *Ann. N.Y. Acad. Sci. 333*:16, (1980).

W.S. Knowles, *Acc. Chem. Res. 16*:106–112 (1983).

W.S. Knowles and M.J.Sobacky *J. Chem Soc.*, Chem. Commun. 1445 (1968).

W.S. Knowles, M.J. Sobacky, and B.B. Vinegard, *J. Chem. Soc.*, Chem,. Commun. 10 (1972).

C.R. Lander, J. Halpen. *J. Organomet. Chem. 250*:485 (1983). Asymetric Synthesis, J.D. Morrison Ed., (Academic Press, San Diego, CA 1985), Vol. 5, p. 41.

Low Temperature Shift Catalyst. U.S. Patent 3,637,528. A. Stiles (1972).

K. Marihara, S. Kawasaki, M. Kofuji, and T. Shimada. Footprint Catalysis. VI. Chiral Molecular Footprint Catalytic Cavities Imprinted by a Chiral Temperature, Bisl N-benzyloxycarbonyl-L-ananyl-amine and their Stereoselective Catalyses. *Bull. Chem. Jpn. 66*:906, (1993).

Methanol Synthesis Catalyst. U.S. Patent 4,111,847. A. Stiles (1973).

Method for the Continuous Preparation of Alcoholates. U.S. Patent 4,327,230. O. Ackermann, H. Leuck, G. Meyer, and G. Schmeling. (1982).

Method for Producing Maleic Anhydride. U.S. Patent 3,906,008. R. Veeda (1975).

Method of Selectively Removing Oxides of Nitrogen from Oxygen-Containing Gases, U.S. Patent 2,975,025. J. Cohn, G.E. Steele, P. Duane, and H.C. Anderson.

Molecular Sieve Adsorbents. U.S. Patent 2,882,243. R.M. Milton (1959).

Molecular Sieve Adsorbents. U.S. Patent 2,882,244. R.M. Milton (1959).

W.R. Moser, D.W. Slocum (eds), *Homogeneous Transition Metal Catalyzed Reaction*. American Chem. Soc., Catalysis Secretariat (Washington, DC, The Soc, 1992). Advances in Chemistry Series, 230.

Novel Reforming Catalysts (ZSM-5). U.S. Patent 4,276,151. C.J. Plank, E.J. Rosinski, and E.N. Givens (1981).

Olefin Polymerization Catalyst. U.S. Patent 3,900,457. D.R. Witt (1975).

Olefin Polymerization and Catalyst Composition for Use Therein. Brit. Patent 2,058,095 A.

Orito, et al., *J. Syn. Org. Chem. Jpn.* 37:173 (1979), *J. Chem. Soc. Jpn.*, 8:1118–1120 (1979); 10:670–672 (1980); 137–38 (1982).

Oxidation of Alkanes to Maleic Anhydride Using Promoted Vanadium Phosphorus Catalysts. U.S. Patent 3,888,886. L.B. Young, B. Weinstein, and A.T. Jurewicz (1975).

Oxide-supported Vanadium Halide Catalyst Components for Olefin Polymerization. Brit. Patent 2,058,939.

Polymers and Production Thereof. U.S. Patent 2,825,721. J.P. Hogan and R.L. Banks (1958).

Polymer-Supported Reactions in Organic Synthesis, P. Hodge and D.C. Sherrington, John Wiley & Sons, Ltd., New York (1984).

Preparation of Catalysts Useful in the Oxidation of SO_2 Gases. U.S. Patent 3,789,019. A. Stiles (1974).

Preparation of Porous Crystalline Synthetic Material Comprised of

Silicon and Titanium Oxides. U.S. Patent 4,410,501. G. Perego, B. Notari (1983).

Process for Ammoxidation of 1 Propanol. U.S. Patent 4,138,271. W. McClellan and A. Stiles (1979).

Process for the Extraction of Platinum Group Metals. U.S. Patent 4,685,963. J. Saville (1987).

Process for Manufacturing Alcohols, Particularly Linear Saturated Primary Alcohols from Synthesis Gas. U.S. Patent 4,122,110. L.A. Sugier and E. Freund (1978).

Process for the Oxidation of Alkenes to Alkene Oxides. U.S. Patent 3,560,530. A. Stiles (1971).

Process for Preparing Modified Silver Catalysts. U.S. Patent 4,033,903. I.E. Maxwell (1977).

Process for Producing Maleic Anhydride. U.S. Patent 3,907,834. E.C. Milberger, S.R. Dolhyj, and H.F. Hardman (1975).

Production of Methanol. U.S. Patent 4,235,799. T.O. Wentworth and A. Stiles (1980).

Production of Oxygenated Hydrocarbons. U.S. Patent 3,326,956. P. Davies and F.F. Snowdon (1967).

B. Pugin and M. Nfdller, Heterogeneous Catalysis and Fine Chemicals, Elsevier Science Publishers, New York (1993).

J.A. Rabo (ed.), Zeolite Chemistry and Catalysis. (ACS Monograph 171), pp. 616–618.

U.S. Patent 3,415,737. Reforming a Sulfur-free Naphtha with a Platinum Rhenium Catalyst. H.E. Kluksdahl (1968).

Ruthenium-Binap Asymmetric Hydrogenation Catalyst. WO 92/15400. A. Chan, W.O. Sun-Chi.

Ruthenium-Phosphine Complex. U.S. Patent 4,994,590. H. Takaya, K. Mashima, H. Kumobayashi, and N. Sayo (1991).

Ruthenium-Phosphine Complex. U.S. Patent 4,691,037. S. Yoshikawa, M. Saburi, T. Ikariya, U. Ishii, and S. Akutagawa (1987).

J. Saville. Proceedings of the Ninth International Process Metals Institute Conference, New York, NY. "Recovery of PGM by Plasma Arc Smelting" (First Commercial Plant).

J. Saville. SAE Detroit, March 1988. Recovery of PGM (Platinum Group Metals) from Spent Auto Emission Catalysts.

R. Scheffold (ed.), Modern Synthetic Methods, Vol. 5, Springer-Verlag, New York, (1989), p. 16–198.

R. Selke, K. Haupke, and H. Karuse. Asymmetric Hydrogenation by Heterogenized Cationic Rhodium Phosphinite Complexes. *J. Mol. Catal.*, 56:315 (1989).

Small Crystal ZSM-5, as a Catalyst. U.S. Patent 5,240,892. D.J. Klocke (1993).

Steam Activated Catalyst. U.S. Patent 3,257,310. C.J. Plank and E.J. Rosinski (1966).

Synthesis of Maleic Anhydride. U.S. Patent 3,904,653. E.C. Milberger and S.R. Dolhyj (1975).

R. Szostak, Molecular Sieves, Principles of Synthesis and Identification; VanNostrand; Reinhold, New York (1989).

A. Tai and T. Harodi; Tailored Metal Catalysts; Y. Iwasawa (ed.). Reidel Pub. Co., Higham, Massachusetts (1986).

Treatment of Silica (Polymerization). U.S. Patent 4,248,735. M.P. McDaniel and M.B. Welch (1981).

J.T. Wehrli, A. Bicker, D.M. Monti and H.U. Blaser, *J. Mol. Catal.* 49: 195–203 (1989).

Zeolite Catalytic Cracking Catalysts. U.S. Patent 3,929,668. G.V. Nelson, D.J. Youngblood, and J.H. Colvert (1975).

Zeolite Containing Compositions for Catalysts. U.S. Patent 3,945,943. J.W. Ward (1976).

Zeolite ZSM-39. U.S. Patent 4,287,166. F.G. Dwyer and E.E. Jenkins (1981).

Index

Milton Keynes UK
Ingram Content Group UK Ltd.
UKHW020021071024
449327UK00032B/2877